Exploring Pure Maths

The MEW Group

Exploring Pure Maths

The MEW Group

Hodder & Stoughton
LONDON SYDNEY AUCKLAND

British Library Cataloguing-in-Publication Data
Berry, John
Exploring pure mathematics.
I. Title II. The MEW Group
510.1

ISBN 0–340–53159–2

First published 1992

Typeset by Litho Link Ltd, Welshpool, Powys, Wales.
Printed in Great Britain for the educational publishing division of Hodder & Stoughton Ltd, Mill Road, Dunton Green, Sevenoaks, Kent by St Edmundsbury Press Ltd, Bury St Edmunds, Suffolk.

Contents

Edited by John Berry and produced by members of The MEW Group.

Charles Ardagh
John Berry
Alan Davies
Steve Dobbs
Roger Fentem
Ted Graham
Barry Hunt
Peter Hudson
Janet Jagger
Brian Lowe
Claire Rowland
Stewart Townend
Nigel Walkey
John Walton
Roger Whitworth

Introduction

The introduction of investigations and problem solving into the assessment component of GCSE has led to more students of mathematics exploring mathematics 'under their own steam'. It is therefore sensible to introduce investigational work into the sixth form classroom, so that students can develop their skills further. This pack is designed to provide short problems to stimulate discussion of mathematics amongst students and between students and their teachers; and to provide some longer investigations that search more deeply into the students' understanding of the topics in GCE Advanced level mathematics.

Mathematics is a subject to explore in its own right as well as to use in problem solving. The algebra, calculus and trigonometry in GCE Advanced level are essentially mathematical methods and there are several resources available for teachers to show the use of these methods in 'real problem solving situations'. However, investigations in pure mathematics can provide an enjoyable experience of 'playing' with mathematics. They need not be long and difficult proofs, indeed problems that might be thought of as recreational can often challenge and develop the students' grasp of a topic.

Investigational work can be used to develop very important problem solving skills. A list of skills common to many investigations is given below:

extract information
gather information
organise information
summarise information
form patterns
extend and generalise
test the generalisation
explain the generalisation

It is difficult, if not impossible, to teach these skills formally. They are learnt by 'doing' investigations and not by 'watching'. Investigations promote inventiveness of thought and help students to be logical and systematic in their approach to problem solving. Furthermore, work of this type gives students the opportunity to develop both oral and written skills.

The student material is in three parts:
Part 1 consists of 90 problems designed to test students' understanding of some of the basic concepts of mathematics. We have called this section *'What and Why . . ?'*, to reflect the aim of the problems which is to help expose common misunderstandings and thus help students to resolve them. These short problems could be used by students individually, but we would suggest that they are better used if tackled as a group activity because this will encourage discussion. It is through discussing their misunderstandings that students can best resolve them. Each problem may take a few minutes to answer and several could be fitted into a lesson. The 90 problems can be classified into eight loose groupings identified in the table overleaf.

Problem numbers	Area of mathematics
1 – 9	Proofs and fallacies
10 – 15	Trigonometry
16 – 37	Algebra and functions
38 – 61	Limits and calculus
62 – 66	Numerical methods
67 – 72	Vectors
73 – 85	Complex numbers
86 – 90	Matrices

Within each band of problems, the level of difficulty and amount of required background knowledge increases as the problem numbers increase.

Part 2 contains 20 longer investigations. The first few of these investigations are suitable for use with GCSE level students as coursework. This was a deliberate policy when planning the pack. Investigations are not an activity to use at the end of term (or the end of year); they should be an integral part of the learning process. Furthermore students could be encouraged to solve these problems using higher level mathematical skills than were available to them at GCSE.

Part 3 provides some photocopiable 'hints and nudges' for students. These should be given out if a group gets stuck while tackling an investigation. They provide a more structured route through an investigation for those who need it.

The pack is designed to provide material for the whole GCE Advanced level range. Most of the investigations can be tackled with different levels of mathematics so that most students can improve their problem solving skills irrespective of their ability level. For the *'What and Why . . ?'* problems, pre-knowledge of the concept or skill will be required so that the problems can be used as revision or to test conceptual understanding.

Inevitably finding teaching time for using investigations in the sixth form is going to be difficult. Over the next few years GCE Advanced level syllabuses and assessment will develop to incorporate and build upon the good practice of GCSE. However, in the short term until these changes occur, we know that many teachers are looking for investigational work to continue to stimulate their students in the way that some coursework activities have stimulated them. One of the main burdens of investigations at GCSE is the marking of student work, and teachers are inevitably concerned about the burden continuing into Advanced level. However we believe that the educational benefits to students of problem solving and investigations are enormous and that these activities should be encouraged in the sixth form with or without any formal assessment. We hope that this pack will allow teachers and students to enjoy working together on problems and investigations that really challenge their understanding of fundamental mathematical ideas.

The MEW (Mathematics in the Everyday World) Group is a group of mathematics teachers who meet regularly to produce material for the various components of GCE Advanced level in Mathematics. There are many textbooks which contain the theory of pure mathematics, statistics, mechanics and decision mathematics, but few contain tests of conceptual understanding and investigations. One of the aims of the MEW Group is to provide extra resources for teachers of these subjects so as to fill the gap. *Exploring Pure Maths* is the second publication of the group and is intended for all courses of mathematical methods at school, college, university or polytechnic at the GCE Advanced level of mathematics. It is part of a package with *Exploring Mechanics* and *Exploring Statistics*. We hope that you and your students enjoy using all three packs.

The authors are indebted to Sharon Ward for typing the draft manuscript, to our families for releasing us for weekends to develop the materials and to the publishers for encouraging us to continue after the publication of *Exploring Mechanics*.

John Berry
Polytechnic South West

SECTION A

STUDENT MATERIAL

PART 1

What and Why..?

Problem 1

$n > 0 \Rightarrow n^2 > 0$

Is the converse true?

Problem 2

If n is even then n^2 is even.
If n^2 is even then n is even.

Are either of the statements true? Back up your answers with proofs.

Problem 3

If a and b are non-zero real numbers, comment on the following statements:

(a) If $a > b$ then $\frac{1}{a} < \frac{1}{b}$

(b) If $\frac{1}{a} < \frac{1}{b}$ then $a > b$

Problem 4

Is it ever true that $\frac{1}{a+b} = \frac{1}{a} + \frac{1}{b}$?

Problem 5

Which of these identities are true and which are false?

(a) $\frac{\log_c a}{\log_c b} = \log_c \left(\frac{a}{b}\right)$

(b) $\log_c \left(\frac{a}{b}\right) = \log_c a - \log_c b$

(c) $\log_c a + \log_c b = \log_c ab$

(d) $\frac{\log_c a}{\log_c b} = \log_c (a-b)$

(e) $\log_c \frac{1}{a} = -\log_c a$

(f) $a\log_c c = a$

(g) $\log_c(-a) = \frac{1}{\log_c a}$

(h) $(c^b c^x)^2 = y$
$\Rightarrow x = \log_c \sqrt{y} - b$

Problem 6

Is it ever true that $\sqrt{a^2 + b^2} = a + b$?

Problem 7

Explain the fallacy in the following argument:

$$\frac{x+3}{x-5} - 4 = \frac{3x-23}{7-x}$$

$$\Rightarrow \frac{x+3-4\,(x-5)}{x-5} = \frac{3x-23}{7-x}$$

$$\Rightarrow \frac{-3x+23}{x-5} = \frac{3x-23}{7-x}$$

$$\Rightarrow \frac{-3x-23}{5-x} = \frac{3x-23}{7-x}$$

Hence $5 = 7$

Problem 8

Explain the fallacy in the following argument:

We know that $3 > 2$ so

$$3\log\frac{1}{2} > 2\log\frac{1}{2}$$

$$\Rightarrow \log\left(\frac{1}{2}\right)^3 > \log\left(\frac{1}{2}\right)^2$$

$$\Rightarrow \left(\frac{1}{2}\right)^3 > \left(\frac{1}{2}\right)^2$$

$$\Rightarrow \frac{1}{8} > \frac{1}{4}$$

$$\therefore\ 1 > 2$$

Problem 9

Explain the fallacy in the following argument:

Suppose that $a = b$

Then $a^2 = ab$

Hence $a^2 - b^2 = ab - b^2$

$\Rightarrow (a-b)(a+b) = b(a-b)$

$\Rightarrow (a+b) = b$

Now substituting $a = b = 1$, we obtain $2 = 1$.

Problem 10

For what values of k can we solve the following equations?

$\sin x = k, \quad |\sin x| = k, \quad \sin x = |k|$

Problem 11

Given that $\sin x \cos x = k$, what happens if $k > \frac{1}{2}$?

Problem 12

Which of the following are identities, and which are equations that can be solved?

(a) $\sin^2 x + \cos^2 x = 1$

(b) $\cos 2x = 2\cos^2 x - 1$

(c) $(x+y)^2 = x^2 + 2xy + y^2$

(d) $\sin x + \cos x = 1$

(e) $\tan 2x = 2\tan x$

(f) $\sqrt{a^2+b^2} = a + b$

(g) $\sin 2x = 2\sin x$

(h) $x^2 = 3x - 2$

Problem 13

Suggest values for x in radians for which $\sin x \approx x$, $\cos x \approx 1$, $\tan x \approx x$.

Problem 14

Are there any values of x for which the following equations are true statements?

(a) $\sin 2x = 2\sin x$

(b) $\cos 2x = 2\cos x$

(c) $\tan 2x = 2\tan x$

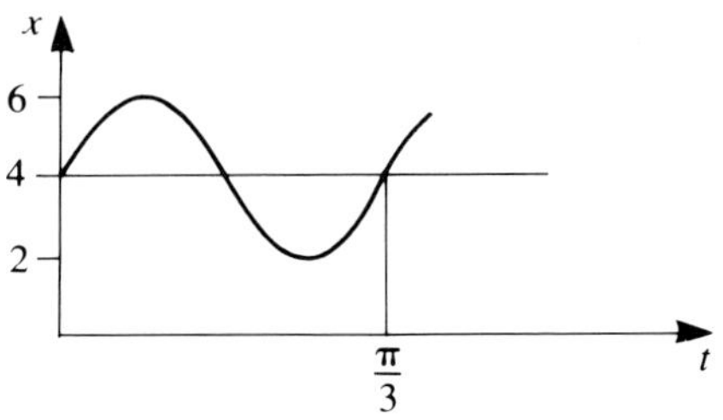

Problem 15

The figure on the left shows a graph of

$$x = a\sin bt + c.$$

What are the values of a, b and c?

Problem 16

What does the word 'linear' mean? Discuss it in the context of the following equations:

(a) $3x + 4y = 7$

(b) $2x - y + 3z = 6$

Problem 17

$$f(x) = ax^2 + bx + c \text{ with } a > 0$$

Under what circumstances will the equation $f(x) = 0$ have

(a) no roots?

(b) one root?

(c) two real roots?

Sketch graphs of $f(x)$ to illustrate the three cases.

Problem 18

Factorise $ax^2 + bx + c$ if the roots of $ax^2 + bx + c = 0$ are α and β.

Problem 19

What can be said about x if

(a) $|x-a| < b$

(b) $x^2 > a$ for $a > 0$

(c) $x^2 > x$?

Problem 20

Give an example of

(a) an equation

(b) an expression

(c) an identity.

Problem 21

Why can $\dfrac{1}{(x-1)(x+2)^2}$ not be rewritten as $\dfrac{A}{(x-1)} + \dfrac{B}{(x+2)}$?

Problem 22

Why can $\dfrac{1}{(x-1)(x^2+1)}$ not be rewritten as $\dfrac{A}{x-1} + \dfrac{B}{x^2+1}$?

Problem 23

What is k if $\ln x = k \log x$? What does this tell you about the graphs of $y = \ln x$ and $y = \log x$?

Problem 24

How is the graph of $y = a^x$ related to the graph of $y = b^x$ where $a, b > 0$? Can you prove this algebraically?

Problem 25

The formula $I = I_0\left(1 + \dfrac{r}{100}\right)^t$ gives the value of an investment after a period of t years.

(a) Explain what I_0 and r represent.

(b) Sketch the graph of I against t.

(c) How does the role of the letters I and t differ from that of I_0 and r in the equation?

Problem 26

If $n! = n \times (n-1) \times (n-2) \times \ldots \times 1$ which has n factors, and $2! = 2 \times 1$ which has 2 factors, and $1! = 1$ which has 1 factor, how can we justify the convention that $0! = 1$?

Problem 27

Explain why

(a) $a(a-4) = b(b-4) \not\Rightarrow a = b$ or $a = b = 4$

(b) $a^2 < b^2 \not\Rightarrow a < b$

(c) $a < b \not\Rightarrow \frac{1}{a} > \frac{1}{b}$

Problem 28

In a plane any two lines must either be parallel or intersect at some point.

(a) What happens to any three lines in a plane?

(b) What can happen to any two lines in three dimensional space?

(c) What can happen to any two planes in three dimensional space?

(d) What can happen to any three planes in three dimensional space?

Problem 29

Write down a function with a modulus sign in it, the graph of which never goes above the x axis.

Problem 30

If transformation A maps the graph of $y = f(x)$ onto the graph of $y = f(2x)$ and transformation B maps the graph of $y = f(x)$ onto that of $y = f(x+\pi)$, why does the graph of $y = \sin(2x+\pi)$ not look like $y = \sin x$ under the transformation BA?

Problem 31

Why are the graphs of $y^2 = x$ and $y = \sqrt{x}$ different?

Problem 32

Sketch the graph of $y^2 = 4x^2$, and comment on your answer.

Problem 33

Can a curve with an equation of the form $y = f(x)$ where f is a function of x ever be symmetrical in the x axis?

Problem 34

Find possible equations for these graphs.

(a)

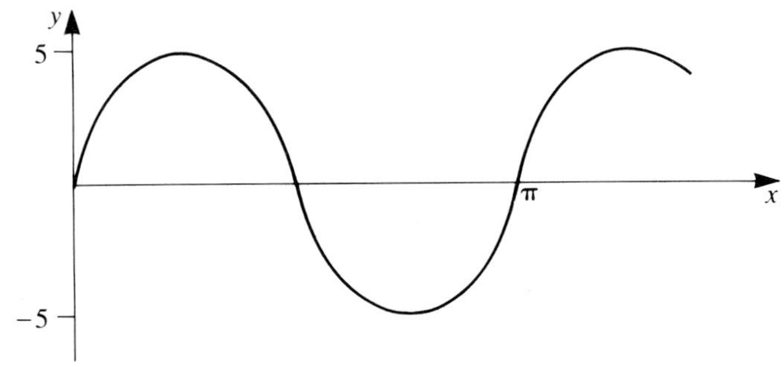

(b)

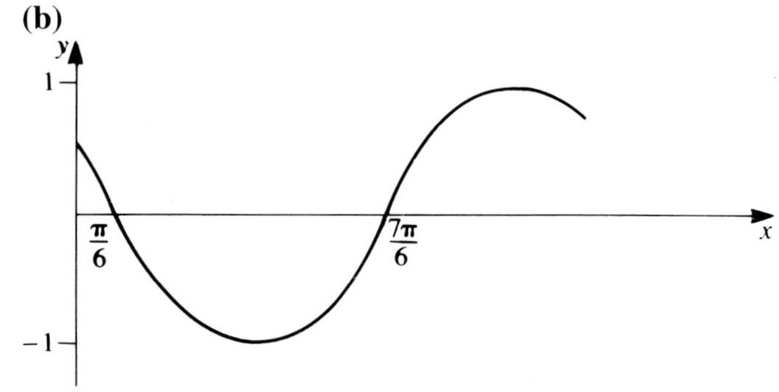

(c)

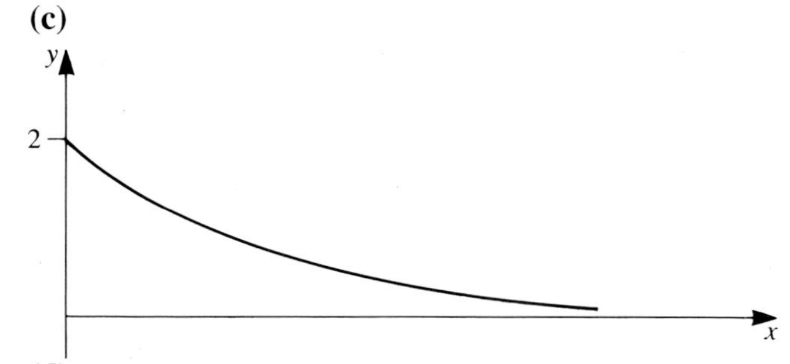

(d)

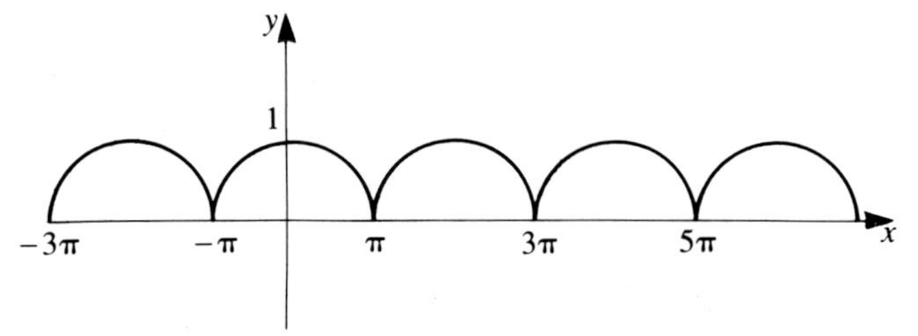

(e)

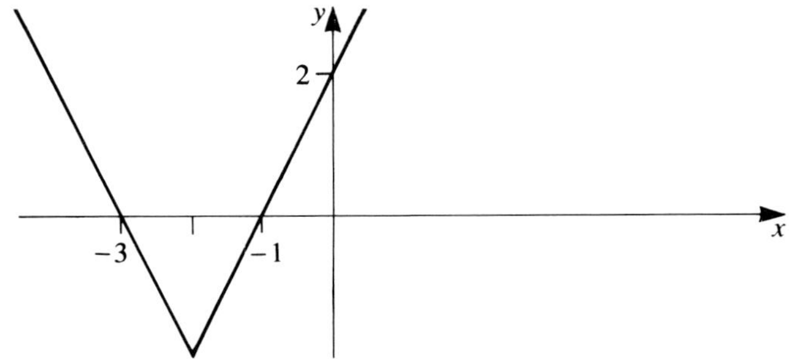

(f)

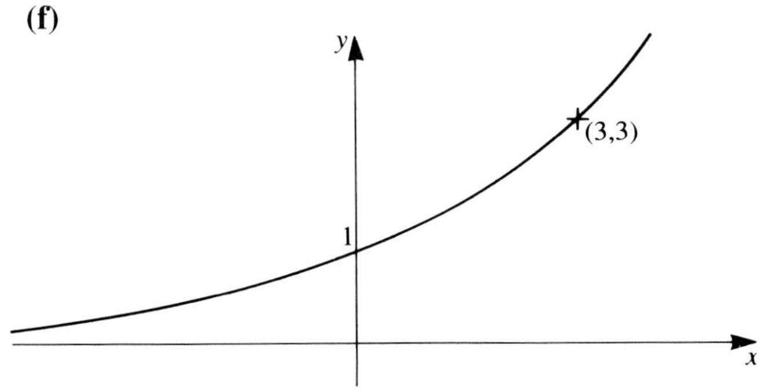

(g)

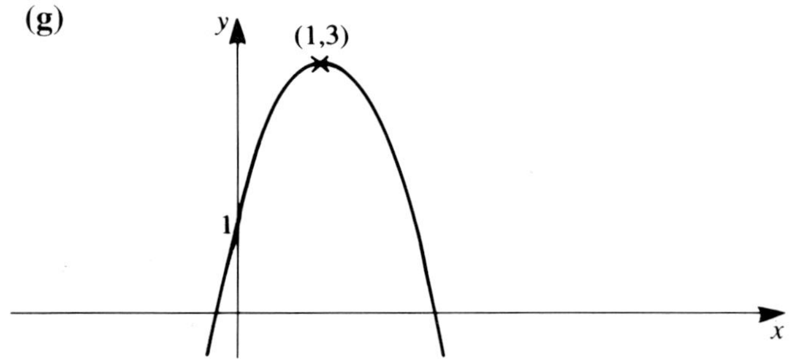

(h)

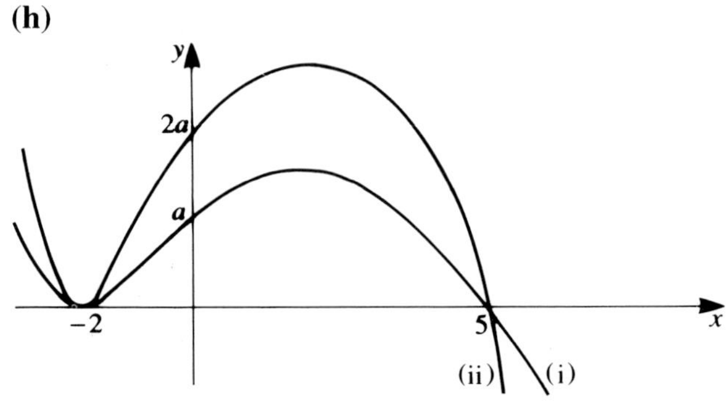

Problem 35

What does the r represent in the polar equation $r = \cos\theta$? Should we allow r to be negative? How does this affect the curve $r = \cos\theta$?

Problem 36

The function $f(x)$ has the property that $f(\alpha) = 0$. Write down a factor of $f(x)$. If, in addition, $f'(\alpha) = 0$ what else can be deduced?

Problem 37

If $P(x)$ is divided by $(x-a)$ the remainder is $P(a)$. What is the remainder if $P(x)$ is divided by (x^2-a^2)?

Problem 38

What happens to $\dfrac{x^2-1}{x-1}$ as x approaches 1, and to $\dfrac{x}{x-2}$ as x approaches 2?

Problem 39

Funny things can happen to a function f(x) when x gets large. As x gets large, what happens to

(a) $\dfrac{3x+1}{5x-4}$

(b) $\dfrac{e^x}{x^{100}}$

(c) $\dfrac{100x+1}{x^2-1}$?

Problem 40

Use your calculator to discover what happens to $\dfrac{\sin\theta}{\theta}$ as θ approaches 0.

Does it matter whether θ is measured in radians or degrees?

Can you deduce what happens to $\dfrac{\sin 2x}{x}$ as x approaches 0 (x in radians)?

Problem 41

Investigate $\frac{1}{2+h} - \frac{1}{2}$ as $h \rightarrow 0$.

Deduce $\lim_{h\rightarrow 0} \left(\frac{\frac{1}{x+h} - \frac{1}{x}}{h} \right)$ for any $x \neq 0$.

Problem 42

'If $y = f(x)$ has a maximum value at $x = a$ then $f'(a) = 0$'.

Is the above statement true?

Is the converse of the statement true?

If the statement or its converse is false provide a counter example.

Problem 43

What is a point of inflection? How is one found?

Problem 44

Sketch $y = |2x-1|$.

What is $\frac{dy}{dx}$ if

(a) $x > \frac{1}{2}$

(b) $x < \frac{1}{2}$

(c) $x = \frac{1}{2}$?

Problem 45

Differentiate the following with respect to x:

$e, \quad \pi^2, \quad a^5, \quad \ln 3, \quad x^e, \quad x^2 \ln 2.$

Problem 46

Why, in calculus, are angles measured in radians and e used as the base of logarithms?

Problem 47

Integrate the following with respect to x:

e^2, x^{π}, ⅓, $2x\ln 3$.

Problem 48

Sean wrote:

$$\int \sin x \cos x \, dx = \frac{\sin^2 x}{2} + c$$

Jane wrote:

$$\int \sin x \cos x \, dx = -\frac{\cos^2 x}{2} + c$$

Can both be correct?

Problem 49

Find the rate of change of the function $f(x) = \ln(\ln(\sin x))$. Explore the graph of the function.

Problem 50

Show why the statement $\int_{-1}^{1} \frac{1}{x^2} \, dx = -2$ is false.

Problem 51

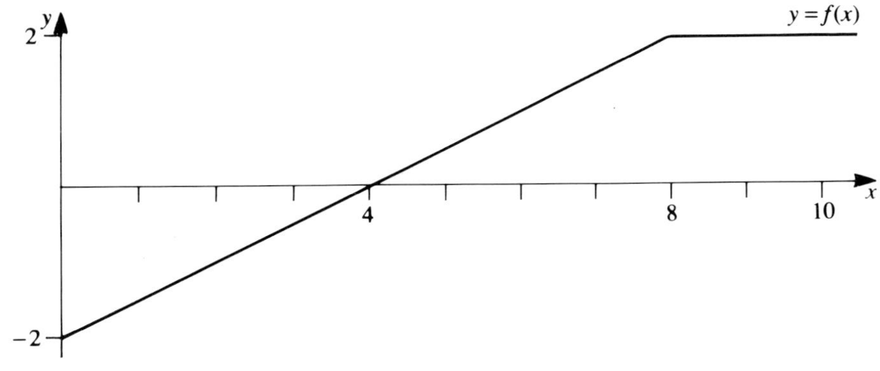

From the figure above find:

(a) $\int_{4}^{10} f(x) \, dx$ (b) $\int_{0}^{4} f(x) \, dx$ (c) $\int_{0}^{10} f(x) \, dx$

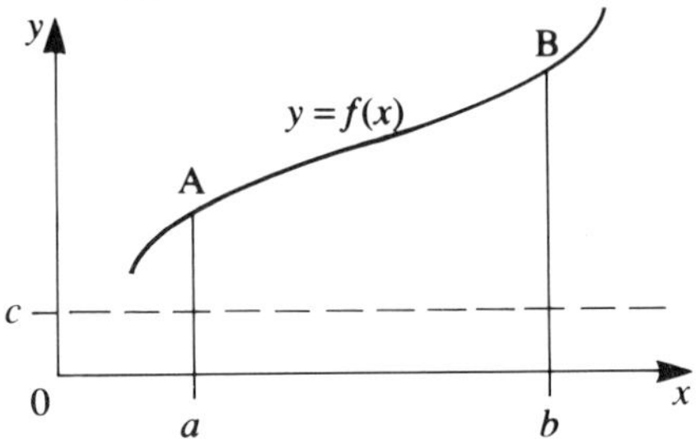

Problem 52

If $\int_a^b f(x)\,dx$ is the area under the curve AB in the figure on the left, state the area under the curve AB which is above the line $y = c$.

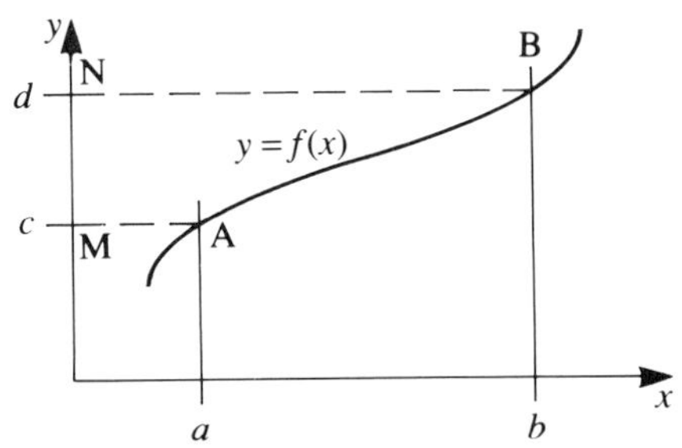

Problem 53

Write down the integral which gives the area between the curve AB and the y axis in the figure on the left, i.e. ABNM.

Problem 54

When is the area A under the curve $y = f(x)$ from a to b equal to

$$\int_a^b f(x)\,dx?$$

What is the connection between area and integral in other cases?

1.

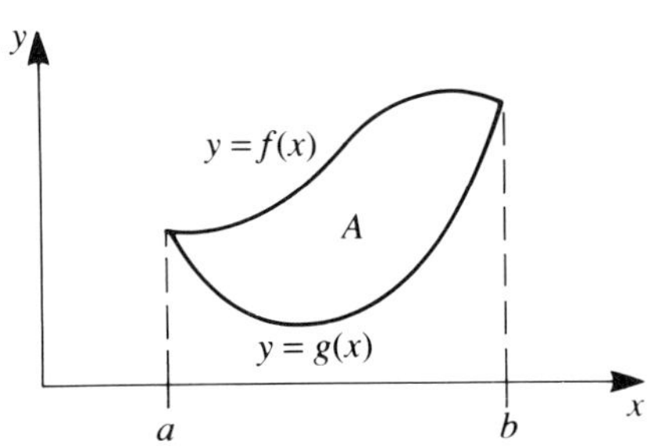

2.

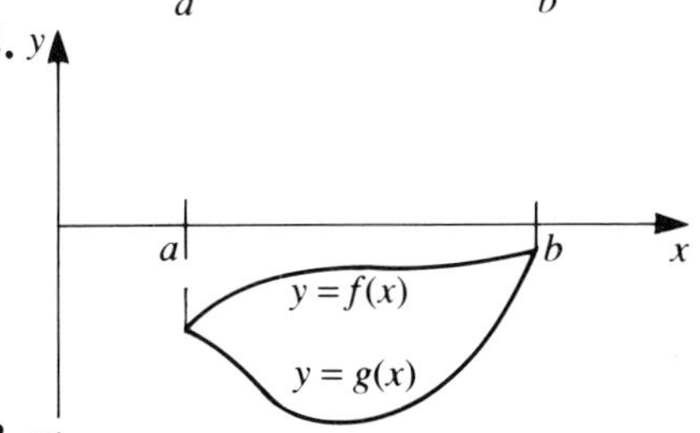

3.

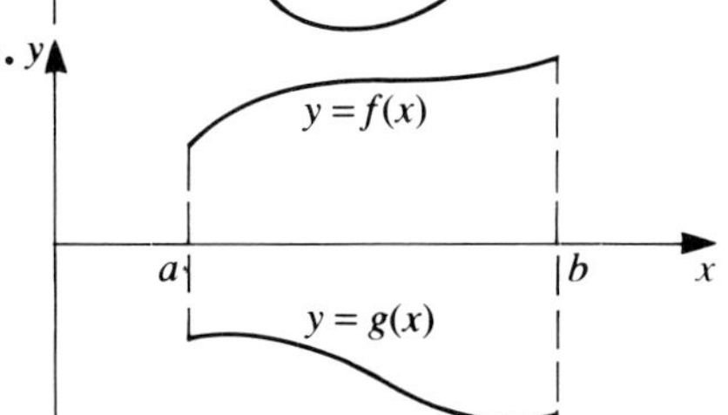

4.

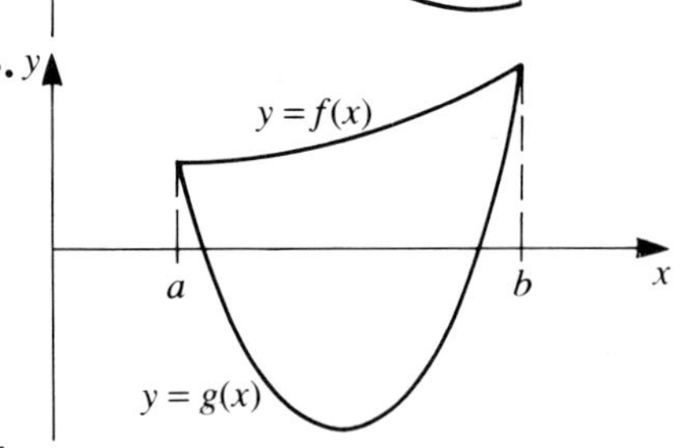

5.

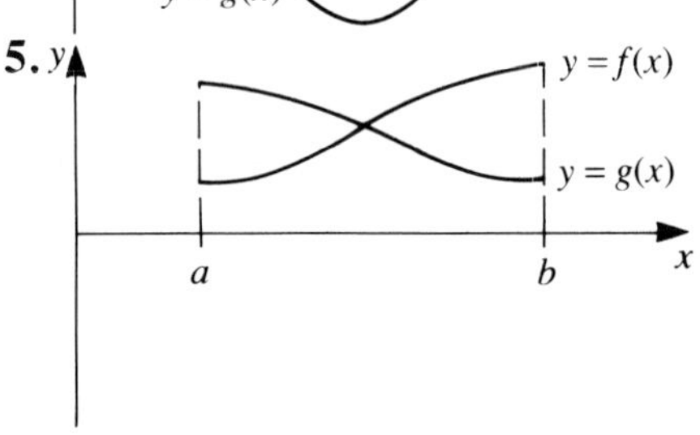

Problem 55

Look at the graphs numbered **1** to **5** on the left. To find the area between two positive curves $y = f(x)$ and $y = g(x)$ where $f(x) > g(x)$ we would say:

$$\text{Area } A = [\text{area under } y = f(x)] - [\text{area under } y = g(x)]$$

$$= \int_a^b f(x)\,dx - \int_a^b g(x)\,dx$$

$$= \int_a^b [f(x) - g(x)]\,dx$$

What is the equivalent result when one, or both, curves are below the x axis, or cross it at some point? What happens if the curves cross each other?

Problem 56

Compare the following curves.

(i) $y = \sqrt{4 - x^2}$ **(ii)** $x^2 + y^2 = 4$ **(iii)** $x = 2\cos\theta$, $y = 2\sin\theta$

In each case, when is the gradient equal to -1?

Problem 57

What must be the form of a first order differential equation so that it can be solved **(i)** by direct integration **(ii)** by separating the variables, **(iii)** by using an approximate numerical method?

Which method(s) can be used on the following equations?

(a) $\dfrac{dy}{dx} = 3x^2 + 2x + 1$

(b) $\dfrac{dy}{dx} = 2x + y,$ given $y = 1$ when $x = 1$

(c) $\dfrac{dy}{dx} = \dfrac{\sin x}{y},$ given $y = 2$ when $x = \dfrac{\pi}{2}$

(d) $\dfrac{dy}{dx} = x^2y$

(e) $\dfrac{dy}{dx} = x^2 + y^2$

(f) $\dfrac{dy}{dx} = \cos 2x,$ given $y = 1$ when $x = 0$

Problem 58

What happens to the integrals

(a) $\displaystyle\int_1^M \frac{1}{x}\,dx$ **(b)** $\displaystyle\int_1^M \frac{1}{x^2}\,dx$

as $M \to \infty$?

Problem 59

If the continuous function $f(x,y)$ has two stationary points which are both maxima then there must be a minimum between them. True or false?

Problem 60

Given that the series $\displaystyle\sum_{n=1}^{\infty} f(x)$ converges, what happens to the integral

$\displaystyle\int_1^M f(x)\,dx$ as $M \to \infty$?

Problem 61

For what values of x does the series

$$\sum_{k=0}^{\infty} \sin^k x$$

converge?

Problem 62

Find $I_1 = \int_0^1 (2x+3)\,dx$, $I_2 = \int_0^1 (3x^2+1)\,dx$ and $I_3 = \int_0^2 4x^3\,dx$

using Simpson's rule with any (even) number of strips.

How accurate is your answer in each case?
Would the trapezium rule give the same answers?

Problem 63

Why is there no error in using Simpson's rule to evaluate

$$\int_{-1}^{2} (5x^3+x^2-3x+1)\,dx?$$

Problem 64

The trapezium rule can be used to give an exact value for the integral of a linear function. Suppose that

$$f(x) = \begin{cases} x & 0 \leqslant x \leqslant 1 \\ 1 & 1 \leqslant x \leqslant 2. \\ 3-x & 2 \leqslant x \leqslant 3 \end{cases}$$

What happens if the trapezium rule is used, with four intervals, to evaluate

$$\int_0^3 f(x)\,dx?$$

Problem 65

The continuous function $f(x)$ has the following properties: $f(-1) = 2$, $f(1) = -1$ and $f'(x) < 0$ for $-1 \leqslant x \leqslant 1$. How many roots of the equation $f(x) = 0$ are there in the interval $[-1, 1]$?

Problem 66

For the function in Problem 65, how many applications of the bisection method are necessary to guarantee that the root will be calculated with a maximum error of less than 0.05?

Problem 67

Show that any vector $\begin{pmatrix} a \\ b \end{pmatrix}$ can be written in terms of the unit vectors $\begin{pmatrix} 1 \\ 0 \end{pmatrix}$ and $\begin{pmatrix} 0 \\ 1 \end{pmatrix}$.

Problem 68

Draw on graph paper the line

$$\mathbf{r} = \begin{pmatrix} 1 \\ 4 \end{pmatrix} + t\begin{pmatrix} 2 \\ -3 \end{pmatrix}$$

Include on your diagram two examples of the vector **r**.

Problem 69

If **a**, **b** and **c** are vectors what can you deduce from the equations

$$\mathbf{a} \cdot \mathbf{c} = 0 \quad \text{and} \quad \mathbf{a} \cdot \mathbf{b} = 0?$$

Problem 70

A triangle is formed by the three vectors **a**, **b** and **c** as shown on the left.

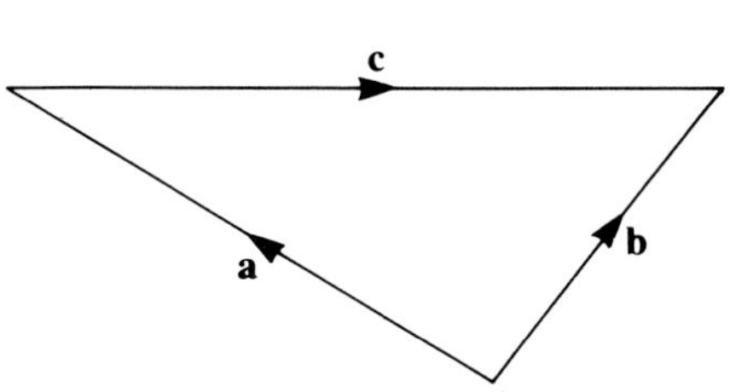

What can you say about the triangle if:

(a) $\mathbf{a} \cdot \mathbf{b} = 0$

(b) $\mathbf{a} \cdot \mathbf{b} < 0$

(c) $\mathbf{a} \cdot \mathbf{c} > 0$

(d) $\mathbf{a} \cdot \mathbf{b} = -\mathbf{a} \cdot \mathbf{c}$

Problem 71

For each of the following problems, in which **a**, **b** and **c** are non-zero vectors, which of the options must necessarily be true?

(a) $\mathbf{a} \cdot \mathbf{b} = 0$

(b) $\mathbf{a} \cdot \mathbf{b} \cdot \mathbf{c} = 0$

(c) $\mathbf{a} + \mathbf{b} + \mathbf{c} = 0$

(d) $\mathbf{a} \cdot (\mathbf{b}\,\mathbf{c}) = 0$

(e) $\mathbf{a} \cdot \mathbf{b} = \mathbf{b} \cdot \mathbf{a}$

Options

(i) The vector **a** is parallel to the vector **b**.
(ii) The vector **a** is perpendicular to the vector **b**.
(iii) The vectors **a, b** and **c** are coplanar.
(iv) Both the vector **a** and the vector **b** are unit vectors.
(v) The relationship is not possible.
(vi) The relationship is true for all non-zero vectors.

Problem 72

In the following expressions **a, b, c** and **d** are arbitrary vectors.

(i)	$\mathbf{a}\cdot(\mathbf{b}\cdot\mathbf{c})$	**(vi)**	$\mathbf{a}\times(\mathbf{b}+\mathbf{c}+\mathbf{d})$
(ii)	$(\mathbf{a}\times\mathbf{b})+\mathbf{c}$	**(vii)**	$(\mathbf{a}\cdot\mathbf{b})-(\mathbf{c}\cdot\mathbf{d})$
(iii)	$\mathbf{a}\times(\mathbf{b}\cdot\mathbf{c})$	**(viii)**	$(\mathbf{a}\cdot\mathbf{b})(\mathbf{c}\times\mathbf{d})$
(iv)	$(\mathbf{a}\times\mathbf{b})\cdot(\mathbf{c}\times\mathbf{d})$	**(ix)**	$(\mathbf{a}\cdot\mathbf{b})(\mathbf{c}\cdot\mathbf{d})$
(v)	$(\mathbf{a}\times\mathbf{b})+(\mathbf{c}\cdot\mathbf{d})$	**(x)**	$(\mathbf{a}\cdot\mathbf{b})\times(\mathbf{c}\cdot\mathbf{d})$

(a) Which of the expressions represent undefined operations?

(b) For which of the allowable expressions is the result a scalar quantity?

(c) For which of the allowable expressions is the result a vector quantity?

Problem 73

If $x = 1 + i$ is a solution of $f(x) = x^3 - 3x^2 + 4x - 2 = 0$, write down the other solutions of $f(x) = 0$.

Problem 74

If $\mathrm{Re}\,(z_1z_2) = \mathrm{Re}\,(z_1)\,\mathrm{Re}(z_2)$ and $\mathrm{Im}(z_1z_2) = \mathrm{Im}(z_1)\,\mathrm{Im}(z_2)$, must both z_1 and z_2 be real?

Problem 75

If $f(\theta) = \cos\theta + i\sin\theta$, which of the following must always be true?

(i) $f(\theta+\Phi) = f(\theta) + f(\Phi)$

(ii) $f(\theta+\Phi) = f(\theta)f(\Phi)$

(iii) $f(-\theta) = -f(\theta)$

(iv) $[f(\theta)]^n = f(n\theta)\ n\varepsilon \mathrm{Z}$.

Problem 76

z is a complex number and a is real, such that $|z-a| = 2$. For the two cases **(a)** $a = 1$ and **(b)** $a = 3$, why can arg(z) in one case take all values and in the other case take only a finite range of values?

Problem 77

The complex number z is such that $|z-2| = 1$. Which of the following must be true for all possible z?

(a) $\mathrm{Re}(z) < 3$

(b) $\mathrm{Re}(z) > 0$

Problem 78

$|z-a| = k$

What does this mean in words?

Problem 79

$\arg(z-a) = k$

What does this mean in words?

Problem 80

Sketch the points represented by $|z| = 2$ in an Argand diagram.

What happens to this sketch if z is replaced by $\dfrac{1}{z}$?

Problem 81

If $z = x + iy$; are the statements

$$|z| = \sqrt{x^2+y^2} \qquad \text{and} \qquad \arg(z) = \tan^{-1} \frac{y}{x}$$

true for all x and y?

Problem 82

What happens to

(a) the line from the origin to the point (3,4) when the coordinates are multiplied by i?

(b) the square with vertices (0,0), (2,0), (2,2), (0,2) when the coordinates are multiplied by i?

(c) the rectangle with vertices (2,1), (5,1), (2,3), (5,3) when the coordinates are multiplied by $\frac{1}{i}$?

Problem 83

What does $|z-3|$ measure?

What can you say about z if

(a) $|z-3| = 2$

(b) $|z+2i| \leqslant \left|\frac{4}{5} + \frac{3}{5}i\right|$

(c) $|z-3| = |z|$?

Problem 84

What does $\arg(z-i)$ measure?

What can you say about z if

(a) $\arg(z-i) = \frac{\pi}{4}$

(b) $|\arg(z-1)| \leqslant \frac{\pi}{3}$

(c) $\arg(z+1) = \arg(z+2i)$?

Problem 85

A real number has zero or two real square roots, one real cube root, zero or two real fourth roots etc.

How many complex roots does a complex number have?

How do they differ?

Problem 86

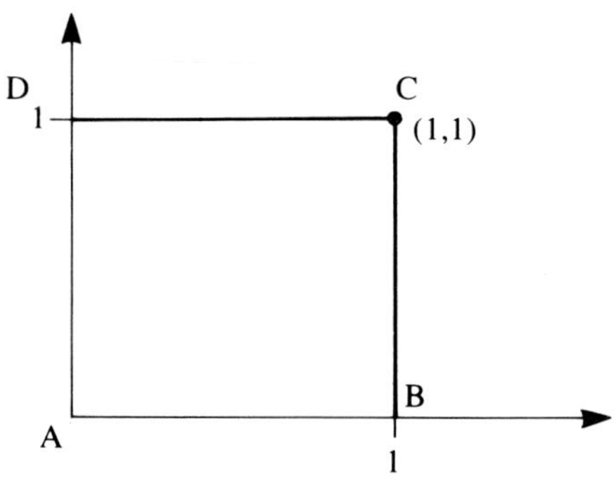

Does every geometric transformation have an inverse?

The transformation represented by the matrix

$$M = \begin{pmatrix} 1 & 1 \\ 1 & 1 \end{pmatrix}$$

is applied to the unit square ABCD shown on the left. The result is that the square becomes A′B′C′D′. M^{-1} does not exist as the matrix M is singular. Describe in words why there can be no geometric transformation which would transform A′B′C′D′ into ABCD.

Problem 87

How could the matrices

$$P = (1\ -1\ \ 2) \qquad Q = \begin{pmatrix} -2 \\ 0 \\ 3 \end{pmatrix} \qquad R = \begin{pmatrix} 4 & 3 & -1 \\ 1 & 2 & 0 \end{pmatrix}$$

be combined to form

(a) a 3×3 matrix

(b) a 2×3 matrix

(c) a 1×2 matrix?

Problem 88

Which geometrical transformation is represented by the matrix

$$J = \begin{pmatrix} 0 & 1 \\ -1 & 0 \end{pmatrix}?$$

Which of the matrices

(i) $\begin{pmatrix} 1 & 0 \\ 0 & 1 \end{pmatrix}$ **(ii)** $\begin{pmatrix} -1 & 0 \\ 0 & -1 \end{pmatrix}$ **(iii)** $\begin{pmatrix} 1 & 0 \\ 0 & -1 \end{pmatrix}$ **(iv)** $\begin{pmatrix} 0 & -1 \\ 1 & 0 \end{pmatrix}$

can be expressed in the form J^n where n is an integer?

Problem 89

Which of the following matrices have a determinant?

(i) (a)

(ii) $\begin{pmatrix} a & b \\ c & d \end{pmatrix}$

(iii) $\begin{pmatrix} a & b & 0 \\ c & d & 0 \end{pmatrix}$

(iv) $(a\,b\,c)$

(v) $\begin{pmatrix} a & b & c \\ a & b & c \\ 0 & 0 & 0 \end{pmatrix}$

Problem 90

If A, B and C are three square matrices of the same size, which of the following statements are not necessarily true?

(i) $(A^{-1})^{-1} = A$

(ii) $(A^{-1})^{T} = (A^{T})^{-1}$

(iii) $\det(AB) = \det(BA)$

(iv) $(A-B)(A+B) = A^2 - B^2$

(v) $\det(A^{-1}) = (\det A)^{-1}$

(vi) If $AB = AC$, then $B = C$

(vii) If $A = B^{-1}$, then $B = A^{-1}$

SECTION A

STUDENT MATERIAL

PART 2

Investigations

Investigation 1: Tower of Hanoi

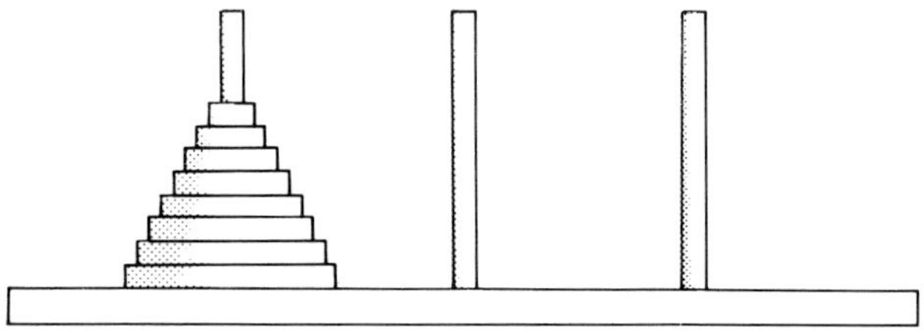

The Tower of Hanoi consists of three pegs fastened to a stand and a number of circular discs of different diameters each of which has a hole in it through which a peg may pass. Initially the discs are all on one peg, the biggest at the bottom and the rest in descending order of size. The object is to transfer the discs to one of the other pegs by moving the discs one at a time from peg to peg in such a way that no disc ever rests on top of a smaller one.

What is the smallest number of moves in which this can be achieved if there are n pegs?

Investigation 2: The cut direct

How many different cross-sectional shapes can be made by cutting a cube?

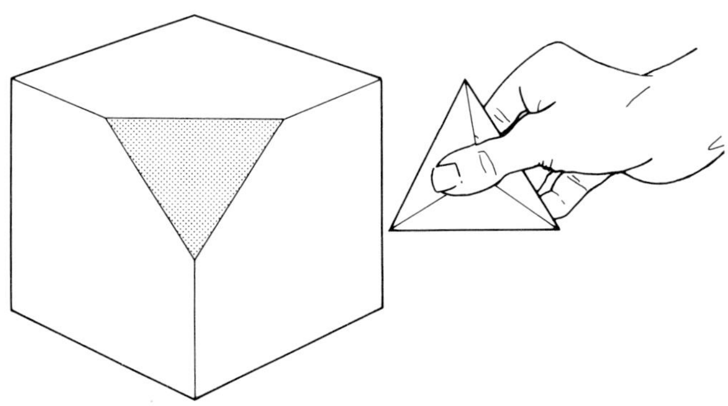

Investigation 3: Cubist paintings

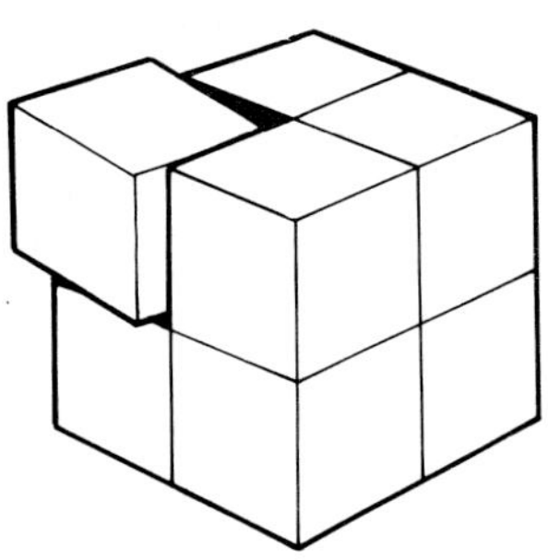

A 2 cm × 2 cm × 2 cm cube is painted pink. It is then cut into 1 cm^3 pieces and the pieces are rearranged so that no pink paint can be seen, and the whole cube is then painted green. How is each 1 cm^3 cube painted?

Is it possible to paint a 3 cm × 3 cm × 3 cm cube pink, green and orange? If so, how is each 1 cm^3 painted? What about a 4 cm × 4 cm × 4 cm cube?

Investigation 4: Patterns with pins and thread

A number, n, of pins is placed at equidistant points on the circumference of a circle. A thread is attached to one pin and then passes round pins at intervals of k steps; this continues until the thread returns to the original pin. For example:

(i) $n = 7, k = 3$ (ii) $n = 6, k = 2$

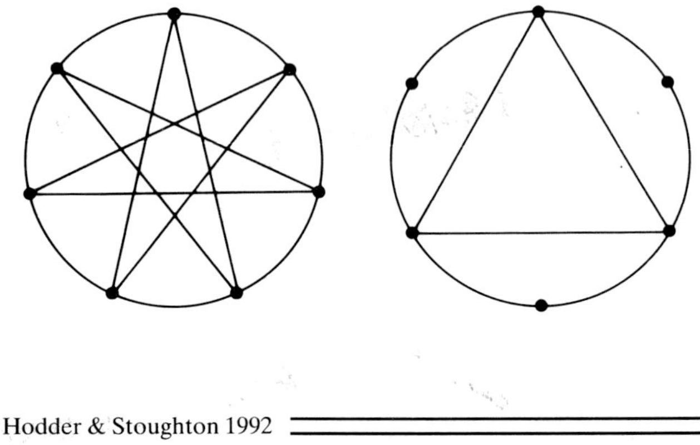

In (i) a complete pattern is obtained, as every point is visited, but not in (ii).

For what pairs of numbers n and k is a complete pattern obtained?

Investigation 5: Sharing a pizza

1 The usual method

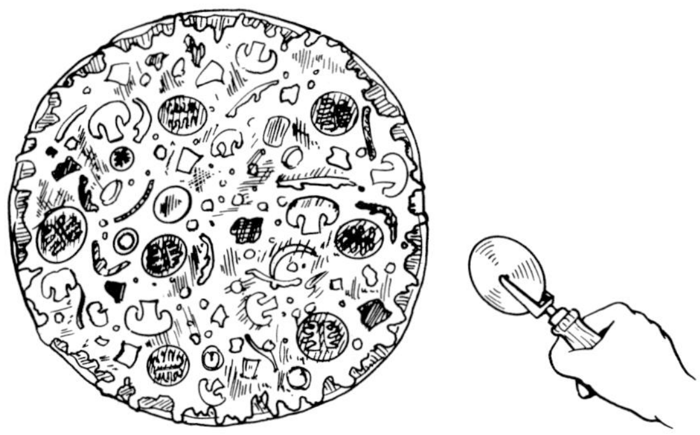

What is the maximum number of pieces into which a circular pizza can be divided by making n cuts?

2 A mathematician's method!

A mathematician decides to cut the pizza in a strange way. She places points on the edge of the pizza and joins each pair of points with a straight cut. For example, 5 points gives 16 pieces of pizza.

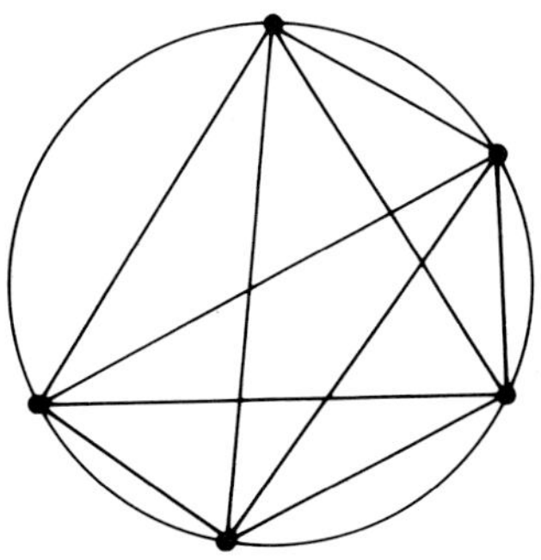

5 points and 16 pieces

What is the maximum number of pieces into which the pizza can be cut by placing n points on the edge and connecting each pair of points by a cut?

Investigation 6: Guess the number

You may have seen a game where the numbers 1 to 64 appear on a set of 6 cards. Each card contains some of the numbers, but a number may appear on more than one card. In the game you are asked to think of a number from 1 to 64 and then indicate on which cards your number is to be found. From this information you will be told the number you were thinking of.

Design a set of six cards with which you would play this game.

Investigation 7: The game of NIM

NIM is a game played by two players with counters.

The counters are placed in a pile and the players take turns to remove counters from the pile; there is a maximum and/or minimum number of counters that can be taken in one turn. The winner is the player who removes the last counter. The game can also be played with more than one pile of counters, in which case players take turns to remove counters from any one pile. The winner is then the player who removes the last counter from the last remaining pile.

Investigate the game and devise a strategy for winning.

Investigation 8: Platonic solids

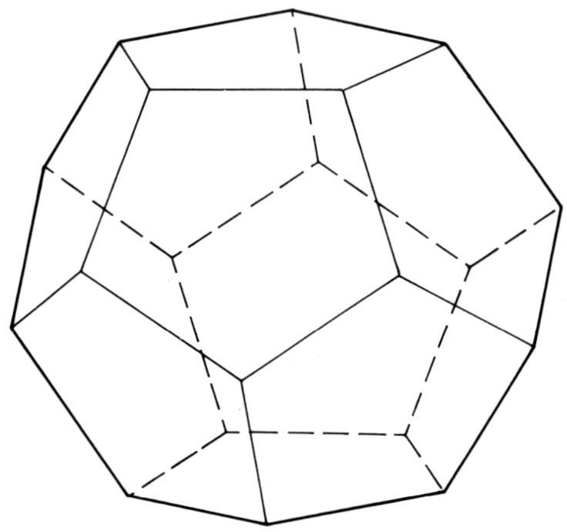

A dodecahedron

Find out the definition of a Platonic solid. Why are there only five Platonic solids and what are they?

Investigation 9: Strange algebra

Investigate solutions of the following pair of equations for the unknowns x and y:

$$20x - 20y = 1$$

$$ax - 20y = k$$

where a and k have fixed values.

Investigation 10: Convergence and divergence

Investigate the convergence of the two series below for various values of p.

(a) $\sum_{k=0}^{\infty} p^k$

(b) $\sum_{k=1}^{\infty} \frac{1}{k^p}$

Investigation 11: Finding roots

Investigate for different values of a the iterative scheme

$$x_{n+1} = ½\left(x_n + \frac{a}{x_n}\right) \quad \text{with } x_0 = 1.$$

Investigation 12: Chaos!

The sequence

$$x_{n+1} = x_n + H(x_n - x_n^2)$$

can be used to model population growth where x_n is the population as a fraction of its equilibrium value. The values $x_0 = 0$ and $x_0 = 1$ are fixed points of the sequence because if $x_0 = 0$ then $s_n = 0$ for all n, and similarly if $x_0 = 1$ then $s_n = 1$ for all n.

Investigate solutions of this sequence for different values of H and x_0.

Investigation 13: Trapezoidal errors grow!

This investigation is concerned with the evaluation of integrals of the form

$$I_r = \int_0^1 x^r e^{-x} dx.$$

Use integration by parts to evaluate I_1, I_2 and I_3.

Develop a recurrence relation involving I_r and I_{r-1} and hence find I_r in terms of r. Check the cases $r = 1, 2$ and 3.

Find I_6 using the recurrence relation, then investigate the evaluation of I_6 using the trapezoidal rule with different numbers of intervals. How does the solution vary with the accuracy of the value taken for e? Try setting up a table of values of I_6 with e taken as 2.7, 2.72, 2.718, etc.

Investigation 14: Rough cuts!

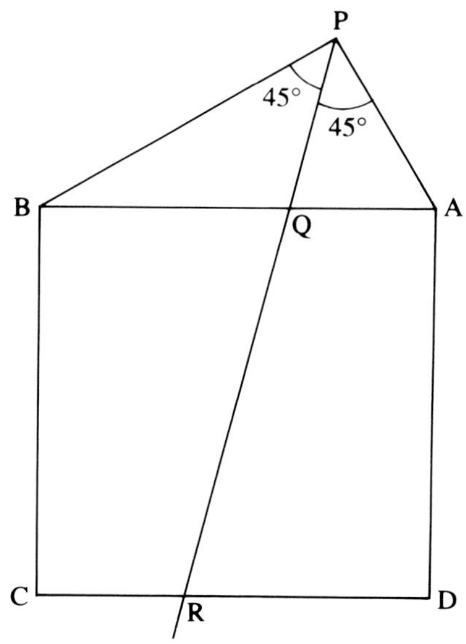

ABCD is a square on the hypotenuse of the right-angled triangle PAB. Line PQR bisects angle P.

How does the line PQR divide the area of the square and the triangle?

Investigation 15: Integer triangle areas

The area of a triangle whose sides have length 13, 14, 15 units is 168 square units. Are there any other triangles whose sides have lengths which are consecutive integers and whose areas are also integers?

Investigation 16: Loci

You will need some sheets of circles like the ones in the figure below. Your teacher can provide photocopied ones.

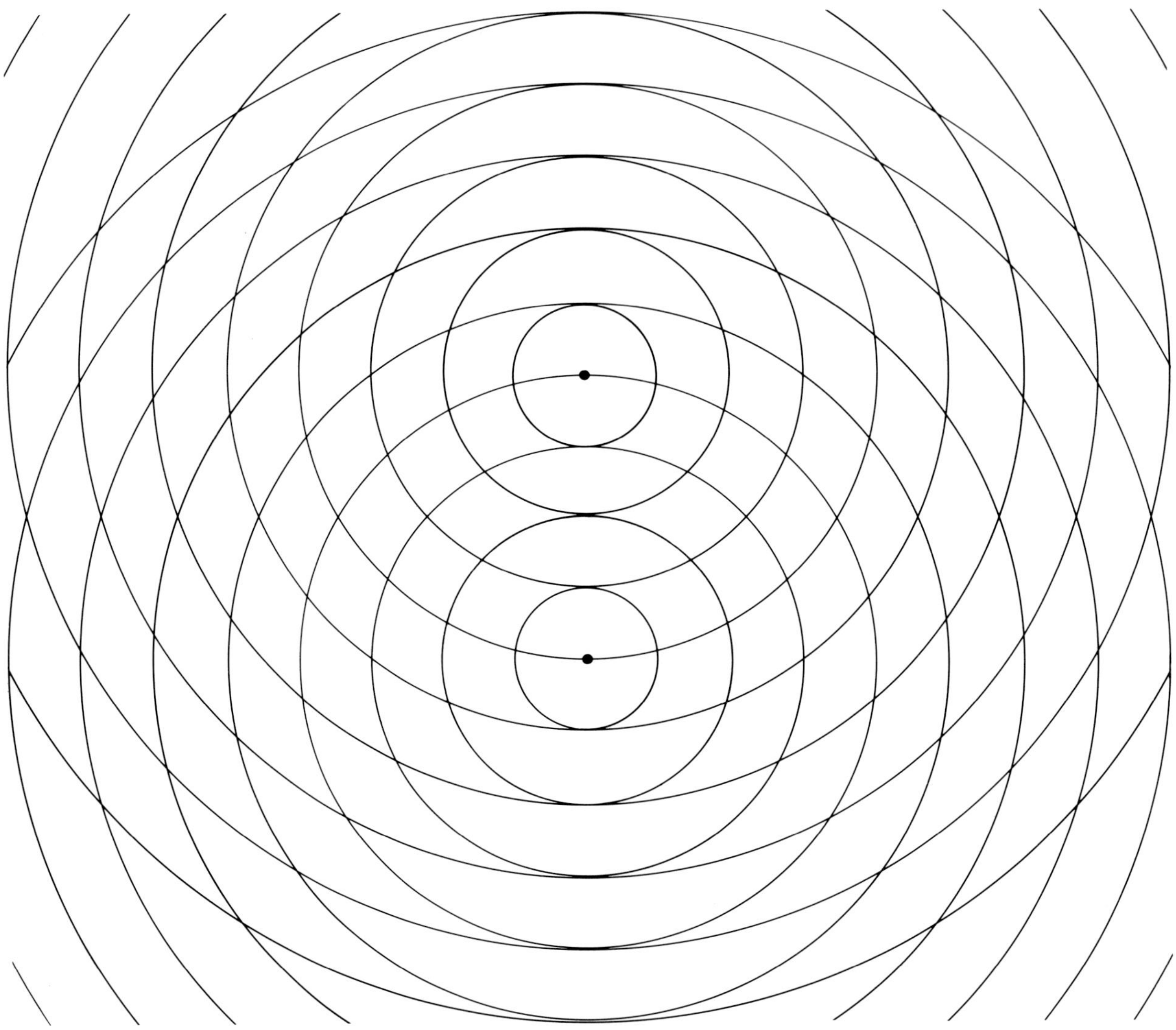

Let A and B represent the two complex numbers a and b on the Argand diagram. Let P represent the complex number z, as in the figure below.

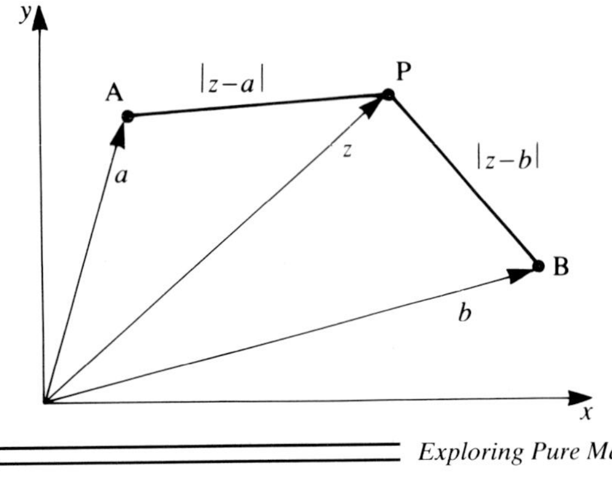

By plotting suitable points for P using the sheets of circles, investigate the loci you can produce with combinations of $|z-a|$ and $|z-b|$. For instance, what is the locus of the point P if $|z-a| = |z-b|$, or if $|z-a| + |z-b| = k$ for some constant k? (You could get started by looking at $|z-a| = k$).

If you draw axes somewhere on the sheet of circles, can you find equations for your loci?

Investigation 17: Snowflakes

The following procedure describes the formulation of the snowflake curve.

Take an equilateral triangle, and on each side, at its midpoint, construct an equilateral triangle whose side is one third the length of the side of the original triangle.

Now repeat this process, constructing an equilateral triangle on each straight line segment of the resulting figure.

If the process were repeated indefinitely, investigate the length of the resulting outline and the area it would enclose.

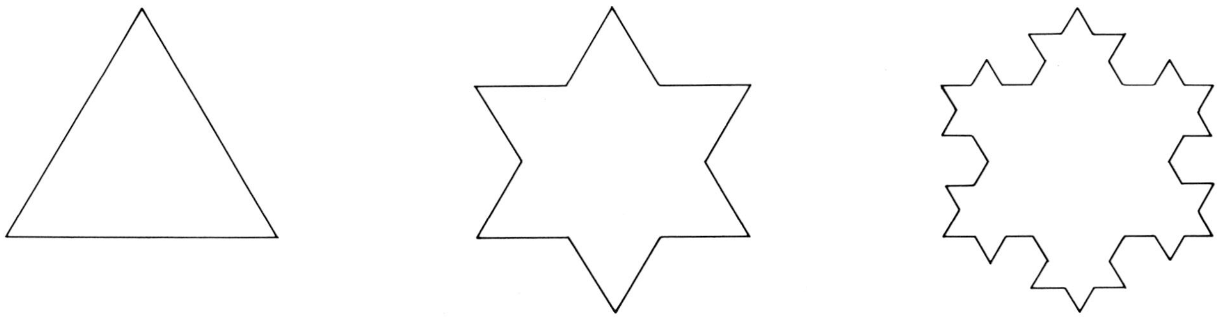

Investigation 18: Into the fourth dimension and beyond

A point (0 dimensions).

Move the point one unit along.

A line (1 dimension).

Move the line through one unit in a direction perpendicular to the line.

A square (2 dimensions).

Move the square through one unit in a direction perpendicular to the square.

A cube (3 dimensions).

Move the cube through one unit in a direction perpendicular to the cube.

A hypercube (4 dimensions).

Investigate!

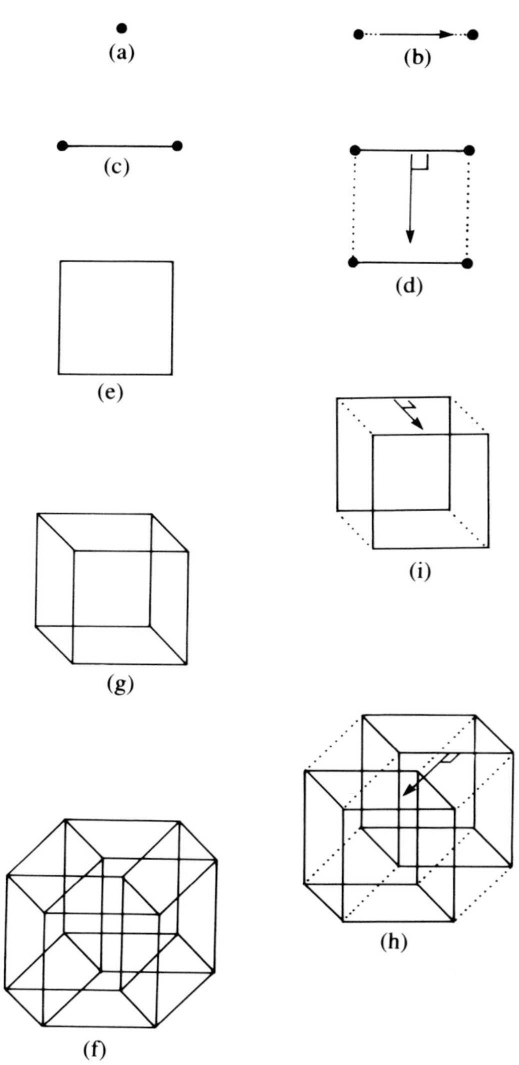

Investigation 19: Continued fractions

A continued fraction is of the form

$$a + \frac{1}{b+x}$$

where x is a continued fraction or a unit fraction.

The notation $[a; \overline{b, c}]$ means $a + \cfrac{1}{b+\cfrac{1}{c}}$.

$[a; \overline{b, c}]$ represents the infinite continued fraction $[a; b, c, b, c, b, c, \ldots]$ i.e.

$$a + \cfrac{1}{b + \cfrac{1}{c + \cfrac{1}{b + \cfrac{1}{c + \ldots}}}}$$

Investigate these types of numbers.

Investigation 20: Functions defined by integrals

Some functions are defined by integrals. Investigate the function $F(x)$ defined by

$$F(x) = \int_0^x f(t)\,dt$$

for a variety of functions $f(t)$.

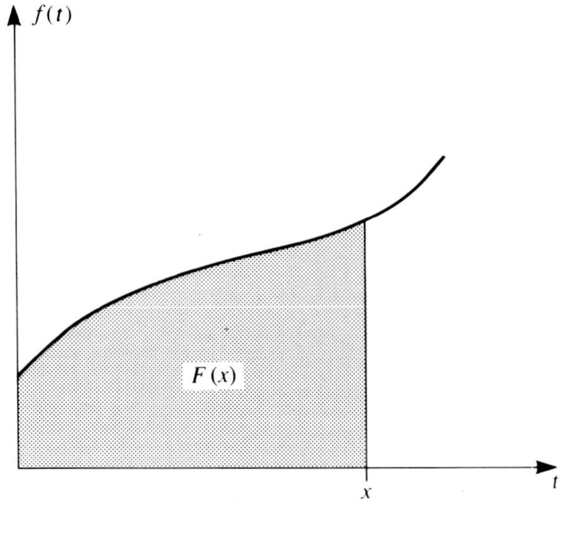

For a function $f(t)$ which does not have a simple primitive function tabulate values of $F(x)$, and plot its graph.

SECTION A

STUDENT MATERIAL

PART 3

Hints and Nudges

Investigation 1: Tower of Hanoi

1 Try it for 1, 2, 3 and 4 discs. Can you generalise?

2 Use trial and error to find a pattern relating the number of moves and number of discs.

3 Find a recurrence relation to generate the number of moves for n discs in terms of the number for $n-1$ discs.

Investigation 2: The cut direct

1 Find some plasticine or soft clay and a sharp old knife.

2 Find the area of each of your shapes.

3 Investigate cutting other solids.

Investigation 3: Cubist paintings

1 Consider the total surface area of all of the separate 1 cm^3 cubes. Is it enough for the different coats of paint on the large cubes?

2 When a large cube is painted in one colour, how many little cubes have three faces painted? How many have two faces painted? How many have one face painted? How many have no faces painted?

Investigation 4: Patterns with pins and thread

1 Try some particular cases, for example $n = 5, k = 2$; $n = 5, k = 7$; $n = 8$, $k = 2, 3, 4$.

2 What happens when $n < k$ or $n = k$? What values of k must be considered?

3 For a complete pattern, which pins must be reached before the original pin is returned to?

4 Try lots of particular cases. What is true about n and k when a complete pattern is not obtained? Conjecture a rule and check other cases.

Investigation 5: Sharing a pizza

The usual method

1 Investigate, by drawing diagrams, how many pieces you get with 1, 2, 3 and 4 cuts. Can you find a pattern? Predict what might happen with 5 and 6 cuts.

2 Try to find a simple rule relating the maximum number of regions and the number of cuts.

A mathematician's method

3 Find the maximum number of pieces for 1, 2, 3 and 4 points on the circle. Can you identify a pattern? What happens with 5 and 6 points? Does it fit your pattern?

4 Try to find a rule. Consider the number of lines joining the points and the number of intersections of these lines. Does this help?

5 The formula relating the number of intersections, I, and the number of points, n, is

$$I = \frac{n(n-1)(n-2)(n-3)}{24}$$

Check that this works. Try to explain why the relationship exists.

Investigation 6: Guess the number

1 The key to the design of the cards is using the binary form of the numbers 1 to 64.

2 What is the binary equivalent of 5, or of 12?

3 Consider the number from 1 to 15 initially — what are the binary equivalents of these, including leading zeros?

4 Group together all numbers whose final binary digit is 1, those whose last but one binary digit is 1, etc. Can you see how to group numbers on to four cards in such a way that we can identify any number just from knowing on which cards it appears?

5 Now consider numbers from 1 to 31 or 1 to 63 and generalise to 1 to 2^n-1, for which n cards will be needed.

Investigation 7: The game of NIM

1 Keep it simple! Use one pile of counters and allow players to take from 1 to 5 (say) counters each turn.

2 Generalise for different maximum (and minimum) numbers of counters removed each turn.

3 What happens if there are two piles of counters, with a player allowed to take any number from one pile? Can you find winning positions?

4 What happens if your opponent also knows the strategy?

5 For three piles why is 1, 2, 3 a winning position? Can you find others?

6 Can you write a BASIC program to play the simplest version of NIM in point 1 above?

Investigation 8: Platonic solids

1 Find some rigid regular polygons and see how they fit together to form a vertex.

2 What is the minimum number of polygons at a vertex?

3 What happens if the sum of the angles at the vertex

(a) equals 360°?

(b) is greater than 360°?

Investigation 9: Strange algebra

The form of the solution depends on the value of k.

(a) Take $k \neq 1$ e.g. $k = 2$.

(i) Consider the four cases $a = -20$, $a = 19$, $a = 21$, $a = 20$.

(ii) Draw graphs to illustrate the results.

(b) Take $k = 1$.

(i) Consider the two cases $a = 20$, $a \neq 20$.

(ii) Draw graphs to illustrate the results.

Investigation 10: Convergence and divergence

1 (a) Consider a specific case, say $p = ½$.

Let $S_n = 1 + \frac{1}{2} + \frac{1}{2^2} + \frac{1}{2^3} + \ldots + \frac{1}{2^n} = \sum_{k=0}^{n} \left(\frac{1}{2}\right)^k$

Use a computer to find the sum to 10, 20, 30, 50 terms and describe the behaviour of the sequence.
What happens as $n \to \infty$?

(b) Repeat for $p = 1.1$. What happens now?

2 (a) Consider the case $p = 1$ (called the harmonic series).

Let $S_n = 1 + \frac{1}{2} + \frac{1}{3} + \frac{1}{4} + \ldots + \frac{1}{n} = \sum_{k=1}^{n} \frac{1}{k}$

Use a computer to find the sum to 10, 100, 1000, 100000 terms and describe the behaviour of the sequence.

(b) Now consider the following series

$$1 + \frac{1}{2} + \frac{1}{4} + \frac{1}{4} + \frac{1}{8} + \frac{1}{8} + \frac{1}{8} + \frac{1}{8} + \ldots$$

Find the sum to 4, 8, 16, 32, 64 terms and describe the behaviour of the sequence. Decide whether or not the harmonic series converges by comparing it to this series. If it converges what is its limit? If it diverges how many terms are required to guarantee that the sum exceeds 10, 100, 1000, 1000000?

(c) Use a computer to investigate the series

$$\sum_{k=1}^{\infty} \frac{1}{k^p}$$

for various values of p, both positive and negative.

Investigation 11: Finding roots

1 Start with $a = 2$ and find the first six terms. Is x_5 the square root of an integer? Can you prove it?

2 Suggest similar iterative formulae for

$\sqrt{3}$, $\sqrt{a}$.

3 What about $\sqrt[3]{a}$ or $\sqrt[p]{a}$?

Investigation 12: Chaos!

1 Start with x_0 between 0 and 1. Investigate solutions for values of H between 0 and 3. What happens to x_n as n gets larger?

2 Here is a simple BASIC program to investigate the solution of the sequence for different values of H and x_0.

```
10   CLS
20   PRINT "What is the value of H?"
30   INPUT H
40   PRINT "What is the starting
     value of X0?"
50   INPUT X0
60   PRINT H;X0
70   FOR N=1 TO 1000
80   XNEW = X0 + H*(X0-X0*X0)
90   PRINT XNEW
100  X0 = XNEW
110  NEXT N
120  END
```

Use it with $x_0 = 0.5$ and $H = 1, 2, 2.5, 3$. Draw graphs of x_n against n. What happens to x_n as n gets larger? Describe the behaviour.

3 Try other values of x_0 and H.

Investigation 14: Rough cuts!

1 Try drawing the figure to scale. Is it possible for R to lie on BC or AD? Try and prove your proposition.

2 Consider R on CD and calculate the ratio. You will probably need to use the sine rule.

Investigation 15: Integer triangle areas

1 Consider right-angled triangles whose sides are pythagorean triples. Work out the areas of these triangles by the simplest method possible.

2 Look at triangles which can be divided into 2 right-angled triangles whose sides are pythagorean triples and calculate the area of these triangles.

3 Look at methods other than ½ base × height for calculating the area of triangles.

Investigation 16: Loci

1 What information is given by $|z-a|$ and how is it represented on the sheet of circles?

2 Try finding the loci given by

$|z-a| = k$

$|z-a| = |z-b|$

$|z-a| = k|z-b|$

$|z-a| + |z-b| = k$

$|z-a| - |z-b| = k$

3 Draw in an x-axis through A and B and a y-axis through the midpoint of AB. Add in a scale so that A and B now have definite coordinates. Let $z = x + yi$. Find the Cartesian equations of the curves you have drawn.

4 Try drawing the axes in a different position.

5 Try generating your equations for any constant k and for $a = a_1 + a_2i, b = b_1 + b_2i$.

Investigation 17: Snowflakes

First it would be advisable to generate a snowflake curve for yourself, and to note how the curve is formed.

1 How many new straight line segments are generated from each straight line segment at each stage of the formation process?

2 What is the length of the new segments at each stage of the formation process? Suppose initially that the side of the original equilateral triangle is one unit.

3 How is the area of each new equilateral triangle related to the area of the original equilateral triangle?

Investigation 18: Into the fourth dimension and beyond

1 Consider how each shape is constructed, and draw up a table like the one below.

	Dimensions	Points	Lines	Squares	Cubes	Hypercubes	etc.
Point	0	1					
Line	1	2	1				
Square	2						
Cube	3						
Hypercube	4						
etc.							

2 Can you describe an n dimensional cube?

3 A triangle can be made from a line by adding a point which is not on the line and joining it to the ends of the line.

A tetrahedron can be made from a triangle by adding a point which is not on the plane of the triangle and joining it to the 3 vertices of the triangle.

Can you continue into 4 dimensions?

Investigation 19: Continued fractions

1 Write $\frac{9}{5}$ as a continued fraction. Write other numbers as continued fractions.

2 What number is represented by [1;1,2,3]? What about [0;2,1]? Experiment with other continued fractions.

3 What number does $[1;\overline{1,2}]$ represent? What about other infinite continued fractions?

Investigation 20: Functions defined by integrals

1 Try simple functions e.g. t, t^2, $\sin t$, $\cos t$, e^t, etc. Check the results by differentiation.

2 Try discontinuous functions

e.g. $f(t) = \begin{cases} 1 & 0 \leqslant t \leqslant 1 \\ 0 & t > 1 \end{cases}$

$f(t) = \begin{cases} t & 0 \leqslant t \leqslant 2 \\ 0 & t > 2 \end{cases}$

3 One function which does not have a simple primitive function is e^{-t^2}. Tabulate the corresponding value of $F(x)$ for $x = 0$ to 3 in steps of 0.1. Plot the graph of $F(x)$.

SECTION B

TEACHER MATERIAL

PART 1

What and Why..?

The following teachers' notes refer to Student Material Part 1,* What and Why . . ?, *to be found on pages 1 to 20 in this book.

Problem 1

No. $n^2 > 0 \not\Rightarrow n > 0$, for example when $n = -1$, $n^2 > 0$.

Problem 2

Both statements are true. The first is easy to prove:

If n even, $n = 2p$
$n^2 = 4p^2 = 2(2p^2)$ hence n^2 even.

The second is not so straightforward. It is easier to show that if n is odd, n^2 is odd:

If n is odd, $n = 2p+1$
$n^2 = 4p^2+4p+1$
$n^2 = 2(2p^2+2p)+1$.

All integers are either even or odd.

If n^2 is even, n cannot be odd, therefore it is even.

Problem 3

True if $a > b > 0$, but not if $a > 0$ and $b < 0$.
For example, if $a = 1$ and $b = -2$ then

$$\frac{1}{a} = 1, \frac{1}{b} = -\frac{1}{2} \text{ and } \frac{1}{a} > \frac{1}{b}.$$

Problem 4

This can only be true if $(a+b)^2 = ab$ (hence $ab > 0$) so

$$a^2 + ab + b^2 = 0$$

a^2, b^2 and ab are all positive, so their sum can only be zero if $a = b = 0$.

But a and b must both be non zero numbers for

$\frac{1}{a} + \frac{1}{b}$ to have any meaning.

Thus it is never true that $\frac{1}{a+b} = \frac{1}{a} + \frac{1}{b}$.

Problem 5

(a)	false	**(c)**	true	**(e)**	true	**(g)**	false
(b)	true	**(d)**	false	**(f)**	true	**(h)**	true

Problem 6

The relationship is only true if either $a = 0$ or $b = 0$ (or $a = b = 0$).

Problem 7

The solution is $x = \frac{23}{3}$ since $\frac{3x-23}{5-x} = \frac{3x-23}{7-x}$ only when both are zero.

The original statement was an equality to solve not an identity presumed true!

Problem 8

$$\log½ = -\log2 < 0$$

so the inequality changes when we multiply through by log½.

Problem 9

$(a-b) = 0$ so that we cannot divide by $a - b$.

Problem 10

For $\sin x = k$, solutions exist for $-1 \leqslant k \leqslant 1$.
For $|\sin x| = k$, solutions exist for $0 \leqslant k \leqslant 1$.
For $\sin x = |k|$, solutions exist for $-1 \leqslant k \leqslant 1$.

Similar problems can be devised for cos and tan.

Problem 11

Since $\sin x \cos x = \frac{1}{2}\sin 2x$ and $-1 \leqslant \sin 2x \leqslant 1$ no solutions of $\sin x \cos x = k$ exist for $k > \frac{1}{2}$.

Problem 12

The problem is intended to clarify the difference between an **identity** which is always true for all values of the variables and an **equation** which may be true for some values of the variables. (These are called the solutions of the equation of course!)

Identities are:

(a) $\sin^2 x + \cos^2 x = 1$

(b) $\cos 2x = 2\cos^2 x - 1$

(c) $(x+y)^2 = x^2 + 2xy + y^2$

The other expressions are equations.

Problem 13

The answer will depend on the accuracy chosen. For example, if we work to 2 decimal places then

$\sin x \approx x$ for $|x| < 0.31$

$\cos x \approx 1$ for $|x| < 0.10$

$\tan x \approx x$ for $|x| < 0.24$

Problem 14

(a) Expanding $\sin 2x$ we have:

$$2\sin x \cos x = 2\sin x$$

$$\sin x(\cos x - 1) = 0$$

Solutions of this equation are $x = 180°n$ or $n\pi$ for integer n.

Hence $\sin 2x = 2\sin x$ for $x = 180°n$ or $n\pi$, when each side is zero.

(b) Replacing $\cos 2x$ by $2\cos^2 x - 1$ we have:

$$2\cos^2 x - 2\cos x - 1 = 0.$$

Solving for $\cos x$,

$$\cos x = \frac{2 \pm \sqrt{4+8}}{4} = \frac{1}{2} \pm \frac{\sqrt{3}}{2}$$

$$\cos x = \frac{1-\sqrt{3}}{2} \quad \text{for real values of } x.$$

$$x = 111.5° + 360°n \text{ or } x = 248.5° + 360°n.$$

These are the values of x for which $\cos 2x = 2\cos x$.

(c) Replacing $\tan 2x$ by $\frac{2\tan x}{1-\tan^2 x}$ we have:

$$\frac{2\tan x}{(1-\tan^2 x)} = 2\tan x$$

$$\tan x \frac{\tan^2 x}{(1-\tan^2 x)} = 0$$

Solutions of this equation are $x = 180°n$ or $n\pi$. These are the values of x for which $\tan 2x = 2\tan x$.

Problem 15

The amplitude is 2 so that $a = 2$.
The period is $\pi/3$ so that $b = 6$.
The intercept on the y axis gives $c = 4$.

Problem 16

A linear equation is one which only contains terms of the first degree in x, y, z (or whichever variable is being used), and a constant. It does not necessarily represent a straight line. For example, $2x - y + 3z = 6$ represents a plane.

Problem 17

(a) $b^2 < 4ac$

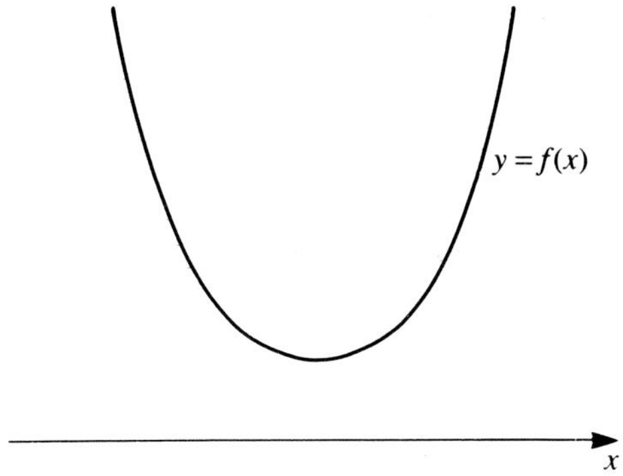

(b) $b^2 = 4ac$

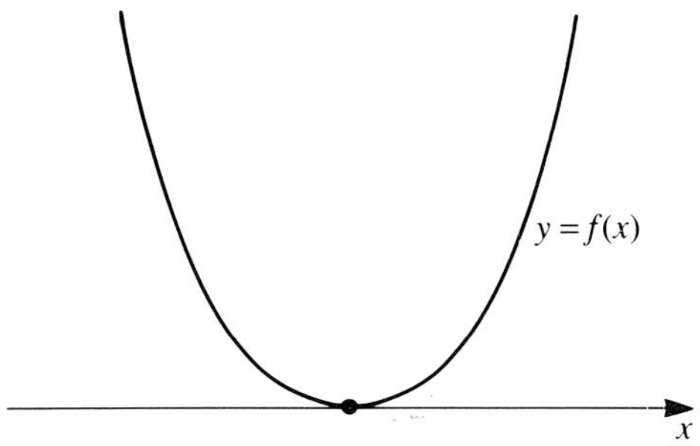

(c) $b^2 > 4ac$

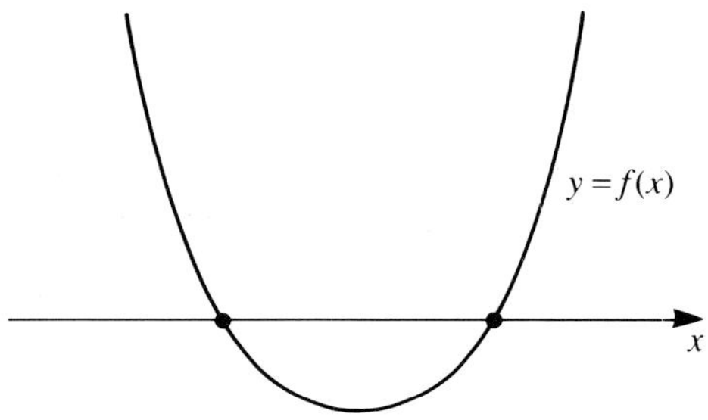

In the figures above we have omitted the y axes, since the situation is the same whether the roots are positive or negative.

Problem 18

If the roots of $ax^2 + bx + c = 0$ are α and β then

$$(x-\alpha)(x-\beta) = 0.$$

Expansion of the brackets gives

$$x^2 - (\alpha+\beta)x + \alpha\beta = 0.$$

Hence we can write

$$ax^2 + bx + c = a(x-\alpha)(x-\beta).$$

Problem 19

(a) $|x-a| < b$

x lies within a distance b of a i.e. $a-b < x < a+b$

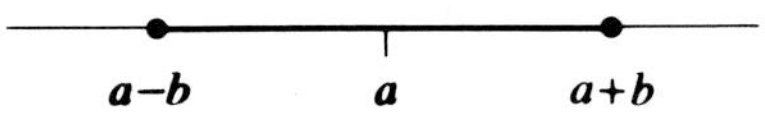

(b) $x^2 > a$

$x^2 - a > 0$

$x < -\sqrt{a}$ or $x > \sqrt{a}$

i.e. $|x| > \sqrt{a}$

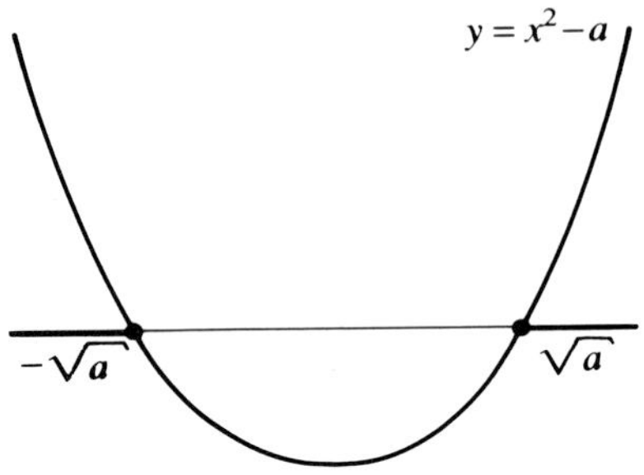

(c) $x^2 > x$

It is always true for $x < 0$ and $x > 1$ but not for $0 \leq x \leq 1$. So $x < 0$ or $x > 1$.

Problem 20

There are many examples that students could give. Knowing the difference between an equation, an expression and an identity is a common difficulty among students (see also Problem 30). Here is one example of each:

(a) $x^2 - 1 = 0$ is an equation,

(b) $x^2 - 1$ is an expression,

(c) $x^2 - 1 = (x-1)(x+1)$ is an identity.

Problem 21

$$\frac{A}{(x-1)} + \frac{B}{(x+2)} = \frac{A(x+2)+B(x-1)}{(x-1)(x+2)}$$

This gives the wrong denominator. To get the right denominator we must multiply top and bottom by $(x+2)$, to give

$$\frac{A(x+2)^2+B(x-1)(x+2)}{(x-1)(x+2)^2}.$$

So $A(x+2)^2 + B(x-1)(x+2) = 1.$

Equating coefficients we get

$$\left.\begin{aligned} A + B &= 0 \\ 4A + B &= 0 \\ 4A - 2B &= 1 \end{aligned}\right\}$$

Now we have three equations but only two unknowns. These three equations are mutually contradictory, and the same happens for most partial fractions of this form.

Problem 22

$$\frac{A}{x-1} + \frac{B}{x^2+1} = \frac{A(x^2+1)+B(x-1)}{(x-1)(x^2+1)}$$

So $A(x^2+1) + B(x-1) = 1$.

Hence $\left.\begin{aligned} A &= 0 \\ B &= 0 \\ A-B &= 1 \end{aligned}\right\}$

Again we have three equations and two unknowns when we equate coefficients. For most partial fractions of this form, the three equations are mutually contradictory.

Problem 23

$$\ln x = k\log x$$

Let $\log x = y$

then $x = 10^y$.

Hence $\ln x = \ln 10^y$

$\ln x = y\ln 10$

$\ln x = \log x \ln 10$ so $k = \ln 10$.

This is a constant so $y = \log x$ can be transformed onto the graph of $y = \ln x$ by a one way stretch scale factor $\ln 10$ parallel to the y axis with the x axis invariant.

Problem 24

If a and b are both greater than 1, then $y = a^x$ is transformed onto $y = b^x$ by a one way stretch parallel to the x axis, while the y axis is invariant (see the figure at the top of the next column). This is the relationship between $y = f(x)$ and $y = f(kx)$. So to prove it we show that $a^x = b^{kx}$ for a constant k.

$$\begin{aligned} a^x &= b^{kx} \\ \log a^x &= \log b^{kx} \\ x\log a &= kx\log b \\ k &= \frac{\log a}{\log b} \text{ which is constant.} \end{aligned}$$

We can also say that the scale factor for mapping $y = b^x$ onto $y = a^x$ is

$$\frac{\log b}{\log a}$$

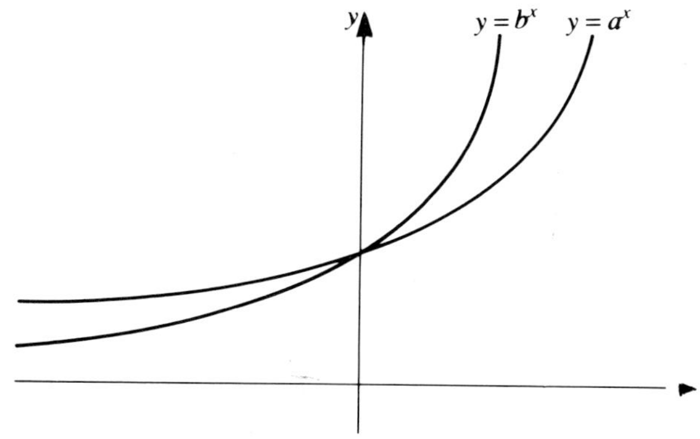

Problem 25

(a) I_0 is the initial investment, r is the percentage rate of interest, interest being added once a year.

(b)

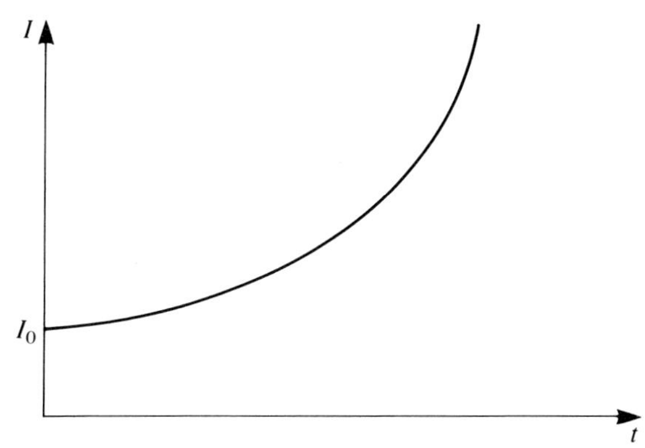

(c) I and t are **variables**, I_0 and r are **parameters**, in other words, whilst they may be varied in order to allow for different investments or interest rates, they will be constant for any specific problem.

Problem 26

$0! = 1$ is a convenient definition, but it is also necessary. Consider for example the following:

${}_nC_r$ is the number of ways of choosing r from n and is equal to

$$\frac{n!}{r!(n-r)!}.$$

In the case where $r = n$, obviously ${}_nC_n = 1$, so

$$\frac{n!}{n!0!} = 1.$$

Thus we need $0! = 1$.

Problem 27

(a)
$$a(a-4) = b(b-4)$$
$$\Rightarrow a^2 - 4a - b^2 + 4b = 0$$
$$\Rightarrow (a-b)(a+b) - 4(a-b) = 0$$
$$\Rightarrow (a-b)(a+b-4) = 0$$

So, whilst $a = b$ is one solution, the other is any set of values for which $a + b = 4$ e.g. $a = 7$, $b = -3$, giving $7 \times (+3) = (-3) \times (-7)$. In general if $ab = cd$, it does *not* follow that $a = c$ and $b = d$ e.g. $2 \times 6 = 3 \times 4$.

(b) If $a = 3$ and $b = -4$, $a^2 < b^2$ but $a > b$.

(c) If $a = -3$, $b = 4$, $a < b$ but $\frac{1}{a} < \frac{1}{b}$.

Problem 28

(a) For three lines in a plane there are various possibilities:
- **(i)** three parallel lines,
- **(ii)** the three lines could form a triangle, with three points of intersection,
- **(iii)** they could all intersect in a point,
- **(iv)** a pair of lines could be parallel, with the third intersecting each line once.

(b) Two lines in three dimensional space may be parallel, intersect at some point or be skew lines.

(c) Two planes must be either parallel or intersect in a line.

(d) For three planes there are various possibilities:
- **(i)** three parallel planes,
- **(ii)** two parallel planes each cut by the third,
- **(iii)** the three planes could intersect in three parallel lines,
- **(iv)** the three planes could intersect along the same line,
- **(v)** they could all meet in a point.

Problem 29

There are many possibilities; one example is $y = -|x|$.

Problem 30

$$\sin(2x+\pi) = \sin\left(2\left(x + \frac{\pi}{2}\right)\right)$$

A is a one way stretch, factor ½, parallel to the x axis with the y axis invariant.

B is a translation $\begin{pmatrix} -\pi \\ 0 \end{pmatrix}$.

But to get from $y = \sin x$ to $y = \sin(2x+\pi)$ you need transformation A followed by the translation

$$\begin{pmatrix} -\pi/2 \\ 0 \end{pmatrix}.$$

$y = \sin(2x + \pi)$ is found after the transformation AB.

Problem 31

For $y^2 = x$, y can take positive or negative values as shown here.

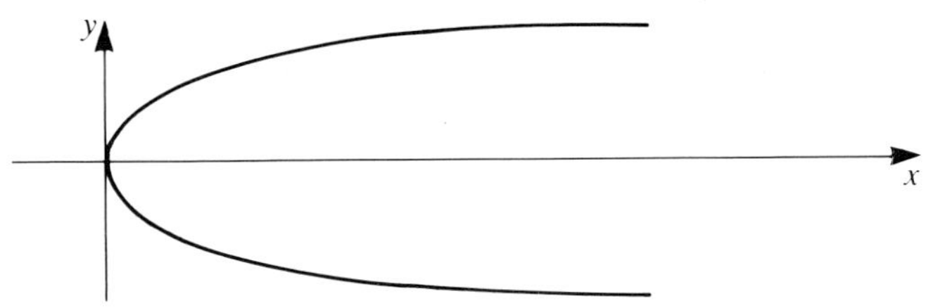

However for $y = \sqrt{x}$, y is defined as taking positive values as can be seen here.

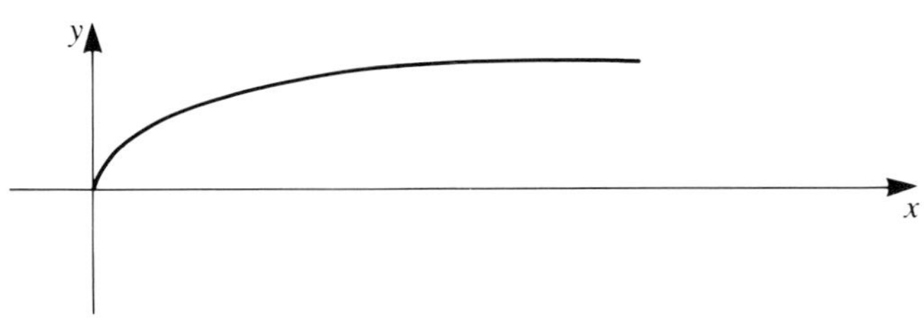

Problem 32

$$\begin{aligned} y^2 = 4x^2 \Rightarrow \quad & y = 2x \\ \text{or} \quad & y = -2x \\ \text{or} \quad & -y = 2x \\ \text{or} \quad & -y = -2x \end{aligned}$$

i.e. $y = 2x$ or $y = -2x$

That is, there are two straight lines, not a curve as you might expect. (See the graph at the top of the next page.)

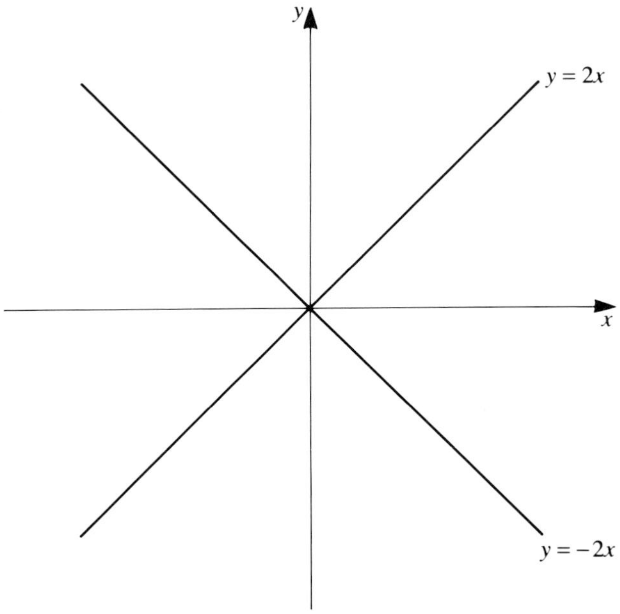

Problem 33

Other than the special case $y = 0$, the answer is no, because for each value of x there is only one value of $y = f(x)$.

Problem 34

There are more possible answers than those listed below!

(a) $y = 5\sin 2x$

(b) $y = \cos\left(x + \frac{\pi}{3}\right)$

(c) $y = 2e^{-x}$ for $x > 0$

(d) $y = \left|\cos\frac{x}{2}\right|$

(e) $y = |2x+4| - 2$

(f) $y = 3^{x/3} \approx (1.44)^x$ $\quad [3 = a^3 \Rightarrow a = 10^{\log 3/3}]$

(g) $y = -2x^2 + 4x + 1$ $\quad [y = -2(x-1)^2+3]$

(h) **(i)** $y = -\frac{a}{20}(x+2)^2(x-5)$

(ii) $y = -\frac{a}{10}(x+2)^2(x-5)$

Problem 35

Although in polar coordinates r is commonly referred to as a distance, some textbook definitions do allow r to take negative values. For example the points P$(2, \pi/4)$ and Q$(-2, \pi/4)$ are shown in the following figures.

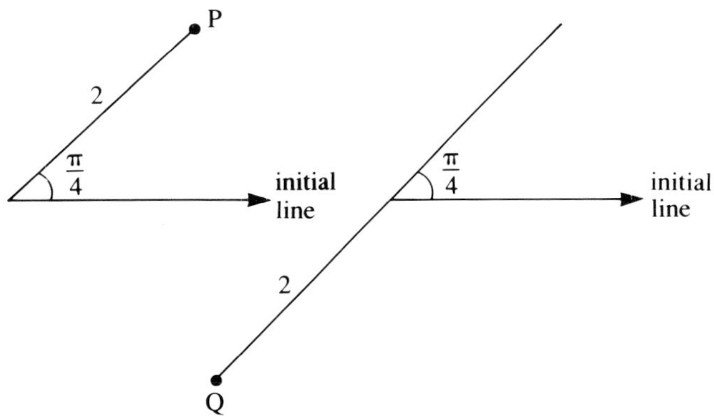

P is the point $(2, \pi/4)$

Q is the point $(-2, \pi/4)$

In fact, $r = \cos\theta$ is the equation of a circle with diameter 1 unit, as explained below.

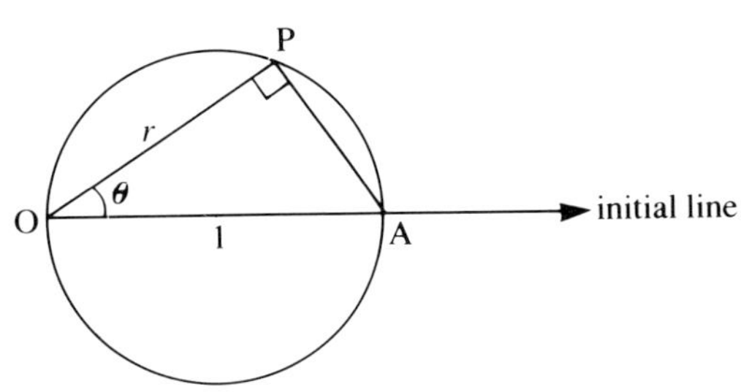

Clearly, in the figure above OA$\cos\theta = r$, and so $r = \cos\theta$ since OA = 1.

For $0 \leqslant \theta \leqslant \pi/2$ the upper half circle is obtained.

For $\pi/2 \leqslant \theta \leqslant \pi$ the lower half circle is obtained since $\cos\theta$ is negative so that r goes in the reverse direction as shown here.

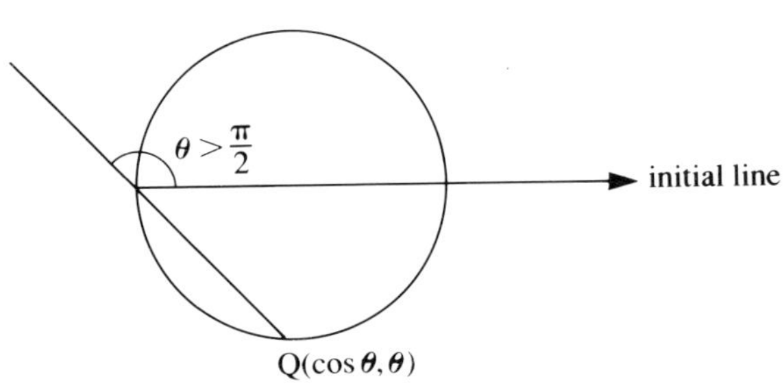

Problem 36

A factor of $f(x)$ is $(x-\alpha)$.

If, also, $f'(\alpha) = 0$ then $(x-\alpha)^2$ is a factor.

Problem 37

The remainder theorem works because

$$P(x) = (x-a)Q(x) + R = (x-a)Q(x)+P(a)$$

Similarly if we divide $P(x)$ by (x^2-a^2) we get a remainder of the form $Rx + S$

$$P(x) = (x^2-a^2)Q(x) + (Rx+S)$$

So $P(a) = Ra + S$

and $P(-a) = -Ra + S$

$\Rightarrow P(a) + P(-a) = 2S$

and $P(a) - P(-a) = 2Ra$

$\therefore \quad S = \dfrac{P(a)+P(-a)}{2}$

and $R = \dfrac{P(a)-P(-a)}{2a}$

So the remainder is

$$\frac{P(a)-P(-a)}{2a}x + \frac{P(a)+P(-a)}{2}$$

Similarly we can tabulate values of $\dfrac{x}{x-2}$ as x approaches 2.

x	$\dfrac{x}{x-2}$
0	0
1	−1
1.9	−19
1.999	−1999
1.999999	−1999999
3	3
2.1	21
2.001	2000
2.000001	2000001

As x approaches 2, $\dfrac{x}{x-2}$ does not approach a single finite number and so $\lim\limits_{x\to 2} \dfrac{x}{x-2}$ does not exist.

Similar problems
Find, where possible

$$\lim_{x\to 2} \frac{x^3-8}{x-2}, \quad \lim_{x\to 1} \frac{x}{(1-x)^2}, \quad \lim_{x\to -3} \frac{(x+3)^2}{2x+7}$$

Problem 38

We can tabulate values of $\dfrac{x^2-1}{x-1}$ for different values of x getting closer to 1, from above and from below.

x	$\dfrac{x^2-1}{x-1}$
0	1
0.5	1.5
0.9	1.9
0.999	1.999
0.999999	1.999999
2	3
1.5	2.5
1.1	2.1
1.001	2.001
1.000	2.000

As x approaches 1, $\dfrac{x^2-1}{x-1}$ approaches 2,

$$\lim_{x\to 1} \frac{x^2-1}{x-1} = 2.$$

Alternatively,

$$\lim_{x\to 1} \frac{x^2-1}{x-1} = \lim_{x\to 1} \frac{(x+1)(x-1)}{(x-1)} = \lim_{x\to 1}(x+1) = 2.$$

Problem 39

There are algebraic approaches to solving these problems, but the following solutions deliberately avoid algebra.

(a) Tabulating x against $\dfrac{3x+1}{5x-4}$ for increasing x

shows that $\lim\limits_{x\to\infty} \dfrac{3x+1}{5x-4} = 0.6.$

Alternatively, when x is large,

$$\frac{3x+1}{5x-4} \text{ is similar to } \frac{3x}{5x} = \frac{3}{5} = 0.6.$$

(b) Similarly, $\lim\limits_{x\to\infty} \dfrac{e^x}{x^{100}}$ is ∞, (i.e. limit does not exist)

is an example of how the exponential function increases faster than any power of x.

(c) Also $\lim\limits_{x\to\infty} \dfrac{100x+1}{x^2-1} = 0$, since when x is large the

fraction behaves like $\dfrac{100}{x}$ which tends to zero.

Similar problems
Find

$$\lim_{x\to\infty} \frac{4x+1}{x^2+1}, \quad \lim_{x\to\infty} \frac{2x^2-1}{6x^2+10}, \quad \lim_{x\to\infty} \frac{x^2+1}{100x+1000},$$

$$\lim_{x\to\infty} \frac{\log x}{x}, \quad \lim_{x\to\infty} \frac{x^3+x^2+x+1}{e^x}$$

Problem 40

θ°	$\frac{\sin\theta^\circ}{\theta^\circ}$
10	0.01736
1	0.01745
0.1	0.017453283
0.001	0.017453200
0.00001	0.017453293
0.000001	0.017453293

$$\lim_{x\to 0} \frac{\sin\theta^\circ}{\theta^\circ} = 0.017453293 \quad (=\pi/180).$$

θ^c	$\frac{\sin\theta^c}{\theta^c}$
1	0.8415
0.1	0.9983
0.01	0.99998332
0.001	0.999999800
0.0001	0.999999995

$$\lim_{\theta\to 0} \frac{\sin\theta}{\theta} = 1$$

$$\lim_{x\to 0} \frac{\sin 2x}{x} = \lim_{x\to 0} 2\,\frac{\sin 2x}{2x}$$

$$= 2 \times 1$$

$= 2$ using above result with $\theta = 2x$.

Similar problems
Find $\lim_{\theta\to 0} \frac{\tan\theta}{\theta}$ when θ is measured in degrees or radians.

Deduce $\lim_{x\to 0} \frac{\tan 2x}{5x}$, and also find $\lim_{x\to 0} \frac{1-\cos x}{x^2}$.

Problem 41

We can tabulate $\frac{1}{2+h} - \frac{1}{2}$ for varying h as $h\to 0$, or argue algebraically:

$$\frac{1}{2+h} - \frac{1}{2} = \frac{2-(2+h)}{2(2+h)} = -\frac{h}{2(2+h)}$$

which tends to 0 as $h\to 0$.

Similarly $\frac{1}{x+h} - \frac{1}{x} = -\frac{h}{(x+h)x}$

$$\text{and so } \lim_{h\to 0} \left(\frac{-\frac{h}{(x+h)x}}{h} \right) = \lim_{h\to 0} \left(\frac{-1}{(x+h)x} \right) = -\frac{1}{x^2}.$$

Similar problems

Find $\lim_{h\to 0} \frac{(2+h)^2-4}{h}$,

$\lim_{h\to 0} \frac{\sqrt{x+h}-\sqrt{x}}{h}$,

$\lim_{h\to 0} \frac{\sin(x+h)-\sin x}{h}$.

Problem 42

In the following we assume that $f(x)$ is continuous.

The statement is false.

This problem should bring out the meaning of 'maximum', 'greatest value' and 'local maximum'.

Counter examples:
(i) $f(x) = x$, $x \in [0, 1]$ has its maximum value at $x = 1$ but $f'(1)$ does not exist, as the figure shows.

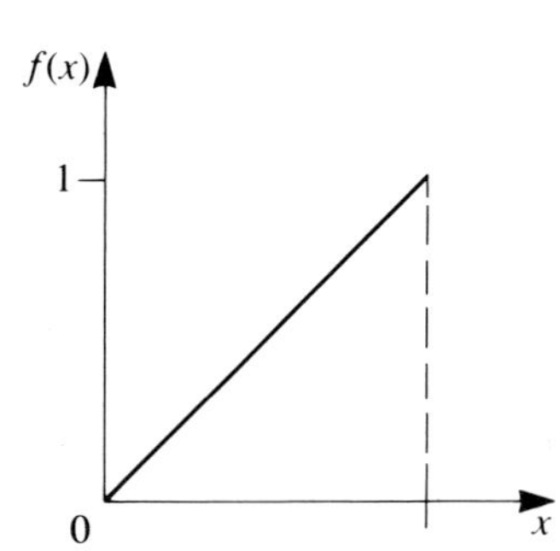

(ii) $f(x) = -|x|$ has its maximum value at $x = 0$, but $f'(0)$ does not exist, as shown here.

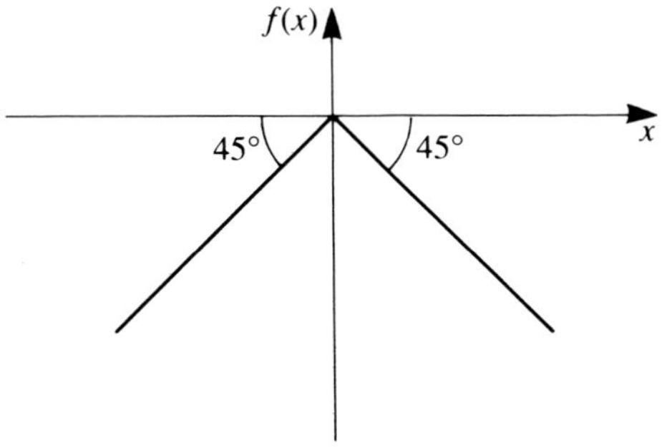

The statement is true for all differentiable functions when a is not an end point of the domain of f.

The converse is

'If $f'(a) = 0$ then $y = f(x)$ has a maximum value at $x = a$.'

This statement is also false.

Counter examples:

(i) $f(x) = x^2$
$f'(0) = 0$ but $x = 0$ is a minimum.

(ii) $f(x) = x^3$
$f'(0) = 0$ but $x = 0$ is neither a maximum nor a minimum.

Similar problems
Are the following statements, or their converses, true? Provide counter examples to false statements.

'If $y = f(x)$ has a minimum value at $x = a$ then $f'(a) = 0$.'

'If $\dfrac{dy}{dx} = 0$ at $x = a$ then there is a maximum or minimum value of y at $x = a$.'

'At a point of inflection $\dfrac{d^2y}{dx^2} = 0$.'

Problem 43

A point of inflection is a point at which a tangent to a curve actually crosses the curve. In other words, it is a point at which the curve changes from concave up to concave down, or vice versa.

The graph of $y = f(x)$ in the following figure has a point of inflection at P. The graph of $y = f'(x)$ has a turning point at the same x-coordinate, so $f''(x) = 0$ at the point of inflection, and is of opposite sign either side of this point.

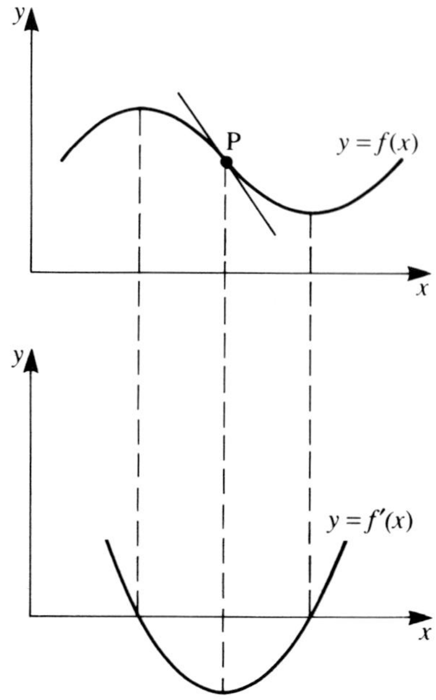

Problem 44

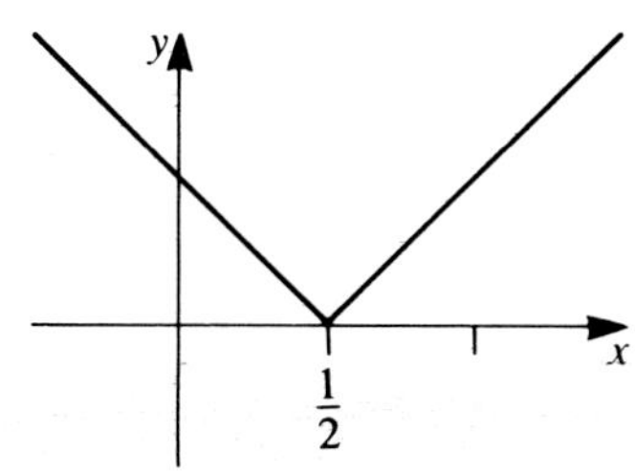

(i) For $x > ½$, $\dfrac{dy}{dx} = 2$

(ii) For $x < ½$, $\dfrac{dy}{dx} = -2$

(iii) When $x = ½$ the derivative is not defined.

Problem 45

Students commonly fail to recognise that e, π^2, a^5, ln3 are constants and will write **wrongly**

$$\frac{d}{dx}(a^5) = 5a^4$$

$$\frac{d}{dx}(\ln 3) = \frac{1}{3}$$

$$\frac{d}{dx}(x^2 \ln 2) = x^2 . \frac{1}{2} \quad -2x \ln 2$$

In fact,

$$\frac{d}{dx}(e) = 0, \qquad \frac{d}{dx}(\pi^2) = 0, \qquad \frac{d}{dx}(a^5) = 0$$

$$\frac{d}{dx}(x^e) = ex^{e-1}, \qquad \frac{d}{dx}(x^2 \ln 2) = 2x \ln 2$$

Similar problems

Differentiate $\frac{1}{\pi}$, xe, x^{π}, $\frac{x^3 \ln \pi}{\sqrt{e}}$

Problem 46

Angles are measured in radians so that the derivatives (and integrals) of the trigonometric functions are as simple as possible. For example,

$$\frac{d}{dx}(\sin x) = \cos x, \text{ when } x \text{ is measured in radians}$$

but $\frac{d}{dx}(\sin x) = \frac{\pi}{180}\cos x$, when x is measured in degrees.

e is used as a base of logarithms since this base gives the simplest results for differentiation and integration. For example,

$$\frac{d}{dx}(\ln x) = \frac{1}{x} \text{ and } \frac{d}{dx}(e^x) = e^x$$

but $\frac{d}{dx}(\log_{10}x) = \frac{1}{x}\log_{10}e$

and $\frac{d}{dx}(10^x) = \frac{10^x}{\log_{10}e}$

Problem 47

Remembering, that e^2, π, etc. are constants:

$$\int e^2\, dx = e^2x + c$$

$$\int x^{\pi}\, dx = \frac{x^{\pi+1}}{\pi+1} + c$$

$$\int \frac{1}{3}\, dx = \frac{1}{3}x + c$$

$$\int 2x \ln 3\, dx = x^2 \ln 3 + c$$

Similar problems

Integrate with respect to x:

$\frac{1}{\pi}$, $2xe$, $x^3\ln 4$, x^{3e}, $\frac{2x\ln\pi}{e^3}$

Problem 48

Both answers are correct as they differ by a constant.

Using Sean's answer, and remembering that $\cos^2x + \sin^2x = 1$,

$$\begin{aligned}\int \sin x \cos x\, dx &= \frac{\sin^2x}{2} + c \\ &= \frac{(1-\cos^2x)}{2} + c \\ &= \frac{-\cos^2x}{2} + \frac{1}{2} + c \\ &= \frac{-\cos^2x}{2} + d\end{aligned}$$

which is Jane's answer with c replaced by d.

Similar problems

$$\int \sec^2x \tan x\, dx = \frac{\tan^2x}{2} + c = \frac{\sec^2x}{2} + d$$

$$\int (1+3x)^2\, dx = \frac{(1+3x)^3}{9} + c = x + 3x^2 + 3x^3 + d$$

Problem 49

This function demonstrates the importance of not rushing headlong into calculus. The rate of change of $f(x)$ is given by

$$\frac{df}{dx} = \frac{\cot x}{\ln(\sin x)}.$$

However, the function $\ln(\ln(\sin x))$ is not defined for any value of x. Sin$x$ lies between -1 and $+1$. To find ln (sin$x$) we can only use values of sin$x$) from 0 to 1. But the value of ln(sin$x$), $0 < \sin x < 1$, will itself be negative and so ln(ln(sin$x$)) does not exist.

It is interesting that a graphics calculator (Casio *fx*-7000G) shows a set of axes on a blank screen. When using graphics calculators and graphical software packages it is important that students understand the properties of functions and can explain what might be unusual behaviour of functions.

Problem 50

The common solution,

$$\int_{-1}^{1} \frac{1}{x^2}\, dx = \left[-\frac{1}{x}\right]_{-1}^{1} = (-1) - (1) = -2$$

is false since the integrand is discontinuous at $x = 0$, which is between the limits of integration.

We must write

$$\int_{-1}^{1} \frac{1}{x^2}\,dx = \lim_{\varepsilon\to 0}\left(\int_{-1}^{-\varepsilon} \frac{1}{x^2}\,dx + \int_{\varepsilon}^{1} \frac{1}{x^2}\,dx\right)$$

$$= \lim_{\varepsilon\to 0}\left(\left[-\frac{1}{x}\right]_{-1}^{-\varepsilon} + \left[-\frac{1}{x}\right]_{\varepsilon}^{1}\right)$$

$$= \lim_{\varepsilon\to 0}\left(\left[\frac{1}{\varepsilon}-1\right] + \left[-1+\frac{1}{\varepsilon}\right]\right)$$

$$= \lim_{\varepsilon\to 0}\left[\frac{2}{\varepsilon} - 2\right] = \infty \text{ (limit does not exist)}$$

$\int_{-1}^{1} \frac{1}{x^2}\,dx$ is an improper integral.

In problems like this, students should be encouraged to explore the properties of the integrand possibly by using a graphics calculator or graphical software package.

Similar problems

$$\int_{-1}^{1} \frac{1}{x}\,dx,\quad \int_{0}^{1} \frac{3}{x^4}\,dx,\quad \int_{1}^{\infty} \frac{1}{\sqrt{x}}\,dx,$$

$$\int_{1}^{\infty} \frac{1}{x^{3/2}}\,dx,\quad \int_{-\infty}^{6} \frac{1}{(4-x)^2}\,dx$$

Problem 51

(a) $\int_{4}^{10} f(x)\,dx = (½ \times 2 \times 4) + (2 \times 2) = 8$

(b) $\int_{0}^{4} f(x)\,dx = (½ \times -2 \times 4) = -4$

(c) $\int_{0}^{10} f(x)\,dx = (-4) + 8 = 4$

Problem 52

$$\text{Area} = \int_{a}^{b} (f(x)-c)\,dx = \int_{a}^{b} f(x)dx - (b-a)c$$

Problem 53

$$\text{Area ABNM} = \int_{c}^{d} x\,dy$$

Problem 54

$$A = \int_{a}^{b} f(x)\,dx$$

provided $f(x) \geqslant 0$ in $[a,b]$.

(i)

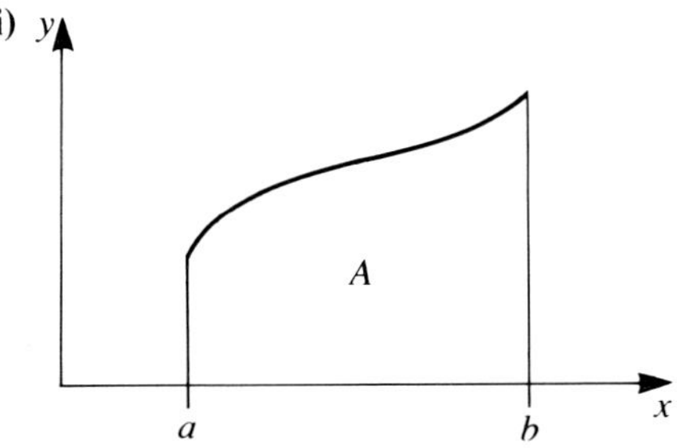

If $f(x)$ is always negative the integral will be negative (why?). In this case

$$A = -\int_{a}^{b} f(x) = \left|\int_{a}^{b} f(x)\,dx\right|$$

(ii)

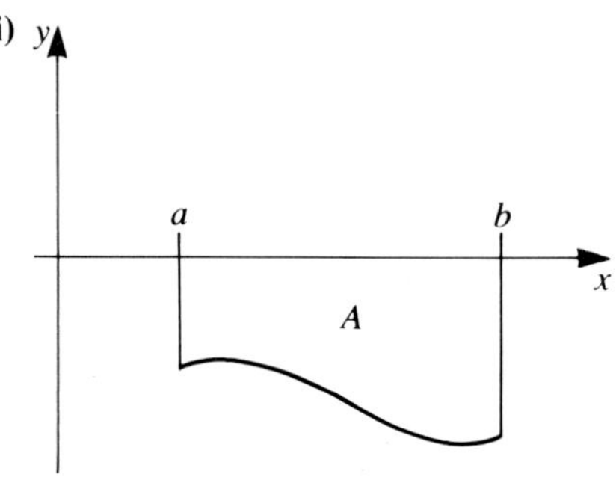

If $f(x)$ is sometimes positive and sometimes negative the parts of the area must be found separately.

$$A = -\int_a^c f(x)\,dx + \int_c^b f(x)\,dx$$

$$A = \left|\int_a^c f(x)\,dx\right| + \int_c^b f(x)\,dx$$

(iii)

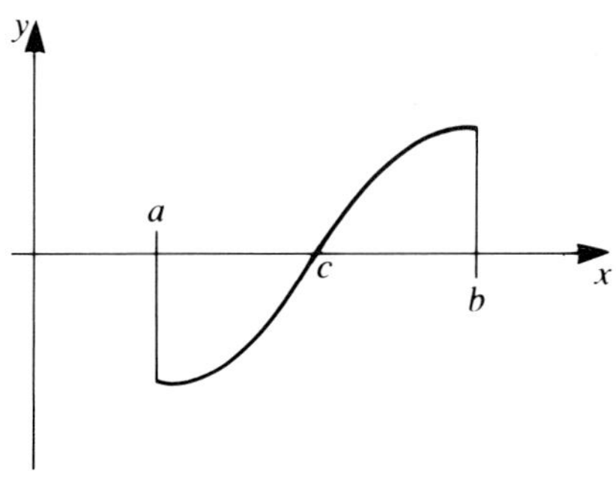

Similar problems

Find the area between the following curves and the x axis.

(a) $y = x^3$ from $x = 1$ to $x = 2$

(b) $y = x^2-9$ from $x = 1$ to $x = 3$

(c) $y = x^2-x-2$ from $x = 0$ to $x = 3$

Problem 55

The result

$$A = \int_a^b \big(f(x)-g(x)\big)\,dx$$

holds in all cases. This can be verified as follows:

Suppose c is a constant so that $y = f(x) + c$ and $y = g(x) + c$ are both positive throughout (a, b). For example, in the figure at the top of the next column,

$$A = \int_a^b \big(f(x)+c\big)\,dx - \int_a^b \big(g(x)+c\big)\,dx$$

$$A = \int_a^b \big(f(x)+c-g(x)-c\big)\,dx$$

$$A = \int_a^b \big(f(x)-g(x)\big)\,dx.$$

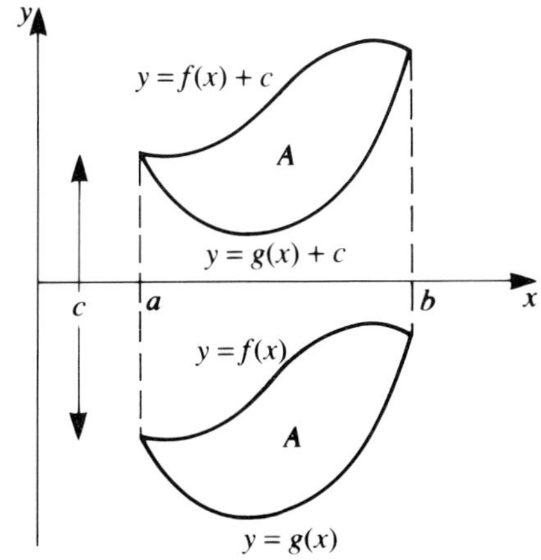

Students could be encouraged to discuss the case when $f(x)$ and $g(x)$ cross, so that $f(x) > g(x)$ for part of the domain and $g(x) > f(x)$ for part of the domain.

Similar problems

Find the areas enclosed by:

(a) $y = x$ and $y = x^2$

(b) $y = x^2-9$ and $y = 2x-6$

(c) $y = 1-x^2$ and $y = x-1$.

Problem 56

(i) $y = \sqrt{4-x^2}$ is the upper half of a circle centre the origin, radius 2, as shown here.

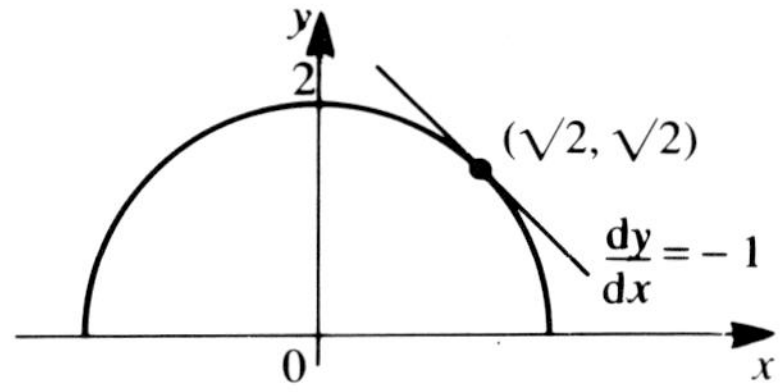

$$\frac{dy}{dx} = \tfrac{1}{2}(4-x^2)^{-1/2}(-2x) = -\frac{x}{\sqrt{4-x^2}}$$

$$\frac{dy}{dx} = -1 \text{ when } x = \sqrt{2}\,, \text{ at the point } (\sqrt{2}\,,\ \sqrt{2}\,).$$

y is given explicitly in terms of x.

(ii) $x^2 + y^2 = 4$ is a circle centre the origin, radius 2, as shown at the top of the next page.

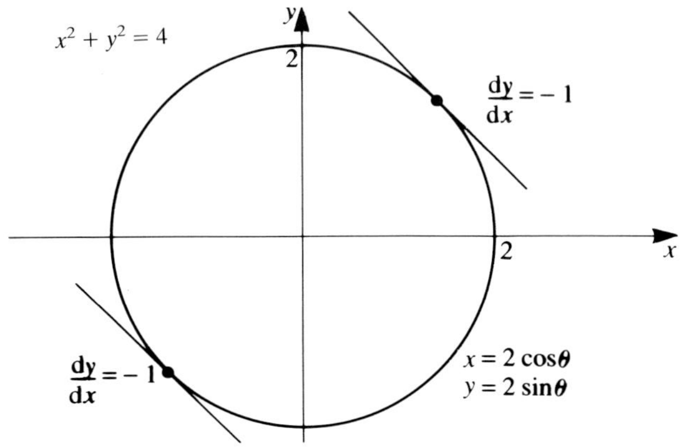

Differentiating: $2x + 2y\dfrac{dy}{dx} = 0$

$$\frac{dy}{dx} = -\frac{x}{y}$$

$$\frac{dy}{dx} = -1 \text{ when } -\frac{x}{y} = -1 \text{ or } x = y$$

The gradient is equal to -1 at $(\sqrt{2}, \sqrt{2})$ and $(-\sqrt{2}, -\sqrt{2})$.

$x^2 + y^2 = 4$ is an implicit equation between x and y.

(iii) $x = 2\cos\theta$, $y = 2\sin\theta$ is also a circle centre the origin, radius 2.

$$\frac{dx}{d\theta} = -2\sin\theta, \quad \frac{dy}{d\theta} = 2\cos\theta$$

$$\text{and } \frac{dy}{dx} = \frac{\dfrac{dy}{d\theta}}{\dfrac{dx}{d\theta}} = -\cot\theta$$

The gradient is equal to -1 at

$$x = 2\cos\frac{\pi}{4}, y = 2\sin\frac{\pi}{4} \text{ i.e. at } (\sqrt{2}, \sqrt{2})$$

and

$$x = 2\cos\frac{5\pi}{4}, y = 2\sin\frac{5\pi}{4} \text{ i.e. at } (-\sqrt{2}, \sqrt{2})$$

$x = 2\cos\theta$, $y = 2\sin\theta$ are parametric equations for x and y.

Similar problems

Consider the curves:

(i) $y = \frac{1}{2}\sqrt{36 - 9x^2}$

(ii) $9x^2 + 4y^2 = 36$

(iii) $x = 2\cos\theta$, $y = 3\sin\theta$

Problem 57

A first order differential equation can be put in the form

$$\frac{dy}{dx} = k(x,y).$$

(i) In order to integrate directly the right hand side can contain terms in x only, and the equation must be of the form

$$\frac{dy}{dx} = f(x).$$

(ii) In order to separate the variables the right hand side must be the product of a term in x and a term in y, and the equation must be of the form

$$\frac{dy}{dx} = g(x)h(y).$$

(iii) In order to solve

$$\frac{dy}{dx} = k(x,y).$$

numerically, an initial condition $y = y_0$ at $x = x_0$ must be given.

	Equation	Direct integration	Separation of variables	Numerical solution
(a)	$\frac{dy}{dx} = 3x^2 + 2x + 1$	✓	✗*	✗
(b)	$\frac{dy}{dx} = 2x + y$, $y=1$ at $x=1$	✗	✗	✓
(c)	$\frac{dy}{dx} = \frac{\sin x}{y}$, $y=2$ at $x = \frac{\pi}{2}$	✗	✓	✓
(d)	$\frac{dy}{dx} = x^2y$	✗	✓	✗
(e)	$\frac{dy}{dx} = x^2 + y^2$	✗	✗	✗
(f)	$\frac{dy}{dx} = \cos 2x$, $y=1$ at $x=0$	✓	✗*	✓

* direct integration could be thought of as a special case of separation of variables; if so these entries would be ✓).

Similar problems
Which method(s) can be used to solve:

$$\frac{dy}{dx} = \frac{1}{1+x^2}, \text{ given } y = 2 \text{ at } x = \frac{\pi}{4}$$

$$\frac{dy}{dx} = \frac{x}{y^3}$$

$$\frac{dy}{dx} = x^3 - \sin y, \text{ given } y = 2 \text{ at } x = 0$$

$$\frac{dy}{dx} = e^x - 1$$

$$\frac{dy}{dx} = \frac{y+1}{e^x}, \text{ given } y = 1 \text{ at } x = 0$$

$$\frac{dy}{dx} = \frac{x+y}{x-y}$$

Problem 58

(a) The integral diverges because $\int_1^M \frac{1}{x}\,dx = \ln M$

and $\lim_{M\to\infty} \ln(M)$ does not exist.

(b) The integral converges because

$$\int_1^M \frac{1}{x^2}\,dx = 1 - \frac{1}{M} \text{ and}$$

$$\lim_{M\to\infty}\left(1 - \frac{1}{M}\right) = 1.$$

Problem 59

False. There could be a saddle point, as shown here.

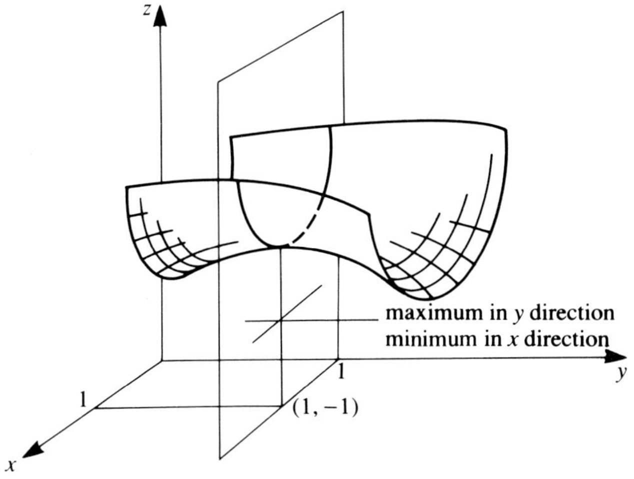

Problem 60

If f(x) is positive and non-increasing for $x > 1$, then the integral converges provided that the series converges. To illustrate the idea, consider

$$\int_1^\infty x^{-p}\,dx \text{ and } \sum_1^\infty x^{-p}$$

which both converge if and only if $p > 1$.

Problem 61

The series converges provided $x \neq (2n+1)\frac{\pi}{2}$.

If $x = (2n+1)\frac{\pi}{2}$ then $|\sin x| = 1$.

Problem 62

With any even number of strips the exact answers $I_1 = 4$, $I_2 = 2$, $I_3 = 16$ are obtained, since Simpson's rule gives exact results for all polynomials of degree less than or equal to 3.

The trapezium rule would give $I_1 = 4$, with any number of strips, since it gives exact results for all linear polynomials, but it would not give I_2 or I_3 exactly. I_2 and I_3 could, however, be found correct to any required number of decimal places by choosing a sufficiently large number of strips.

Similar problems
Show that Simpson's rule with 2 strips gives the exact value of

$$I_1 = \int_0^1 (4-x)\,dx,$$

$$I_2 = \int_1^2 (x^2+x-1)\,dx,$$

$$I_3 = \int_0^1 (x^3-x^2+x-1)\,dx.$$

Problem 63

There is no error in this case because Simpson's rule evaluates integrals of cubic polynomials exactly.

Problem 64

The exact value will be obtained provided the intervals do not span the lines $x = 1$ and $x = 2$.
For instance, [(0, 1), (1, 2), (2, 2.5), (2.5, 3)] will give the exact value.

Problem 65

The function is continuous and always decreases, and the sign changes in the interval; hence there is exactly one root in the interval $[-1, 1]$.

Problem 66

The bisection method halves the maximum error at each step. Five applications are needed.

Problem 67

$$\begin{pmatrix} a \\ b \end{pmatrix} = \begin{pmatrix} a \\ 0 \end{pmatrix} + \begin{pmatrix} 0 \\ b \end{pmatrix} = a\begin{pmatrix} 1 \\ 0 \end{pmatrix} + b\begin{pmatrix} 0 \\ 1 \end{pmatrix}$$

Problem 68

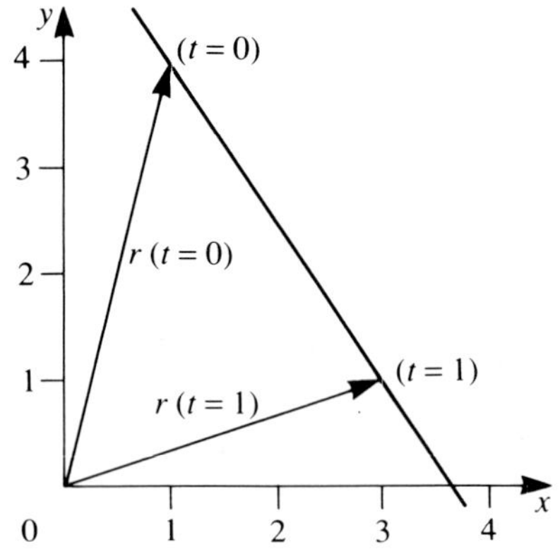

Problem 69

a • **c** = 0 implies **a** = **0**, **c** = **0** or **a** and **c** are perpendicular and similarly for **a** • **b** = 0.

If **a**, **b** and **c** are non-zero vectors in the same plane then **b** and **c** must be parallel since they are each perpendicular to **a**. If **a**, **b**, **c** are general three dimensional vectors then **b** and **c** could be skew lines, both perpendicular to **a**. It is also possible for **b** and **c** to define a plane perpendicular to **a**.

Problem 70

(a) **a** and **b** are perpendicular so we have a right-angled triangle.

(b) The angle between **a** and **b** is obtuse.

(c) If **a** • **c** > 0 then the external angle α is acute and we cannot draw the triangle with **a**, **b** and **c** in the directions shown here.

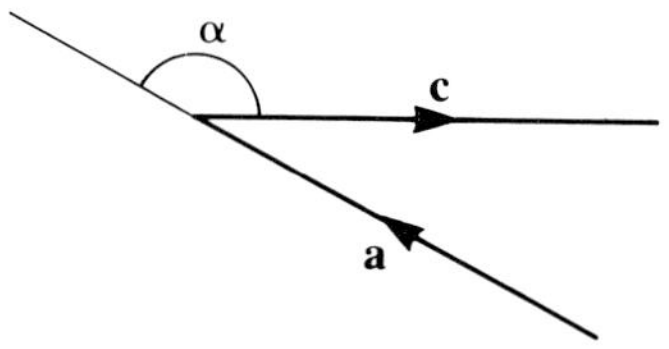

(d) Either **a** = **0** or **b** = −**c**. In either case the triangle does not exist.

Problem 71

(a) Option (ii)

(b) Option (v)

(c) Option (iii)

(d) Option (v)

(e) Option (vi)

Problem 72

(a) Expressions (i), (iii), (v) and (x) are not defined operations.

(b) The scalar quantities are (iv), (vii) and (ix).

(c) The vector quantities are (ii), (vi) and (viii).

Problem 73

If z is a solution of a polynomial equation with real coefficients then so is its complex conjugate, $\bar{z}$. A second solution is therefore $x = 1-i$.

To find the third solution we factorise since $(x-1+i)(x-1-i) = x^2 - 2x + 2$ is a factor.

$$(x^3-3x^2+4x-2) = (x^2-2x+2)\,(x-1)$$

Hence $x = 1$ is the third solution.

Problem 74

If neither z_1 nor z_2 is zero, then z_1 and z_2 must be real. To show the result let $z_1 = a + ib$ and $z_2 = c + id$.

Then $\text{Re}\,(z_1z_2) = ac - bd$ and $\text{Im}(z_1z_2) = ad + bc$

Hence $ac - bd = ac$ so that $bd = 0$

and $ad + bc = bd = 0$

Thus either $b = 0$ or $d = 0$.

If $b = 0$, then $ad = 0$; and either $a = 0$ ($z_1=0$) or $d = 0$ (z_1 and z_2 are both real).
If $d = 0$, then $bc = 0$; and either $b = 0$ (z_1 and z_2 are both real) or $c = 0$ ($z_2=0$).

Problem 75

Statements (ii) and (iv) are true. Using the exponential form of a complex number provides the easiest proof.

$$f(\theta) = e^{i\theta}$$

So (ii) $f(\theta+\Phi) = e^{i(\theta+\Phi)} = e^{i\theta}e^{i\Phi} = f(\theta)f(\Phi)$;

and (iv) $[f(\theta)]^n = e^{in\theta} = f(n\theta)$.

Problem 76

The equation $|z-a| = 2$ represents a circle of radius 2 centre a. For $a = 1$ and $a = 3$ the circles are as shown in the following figures.

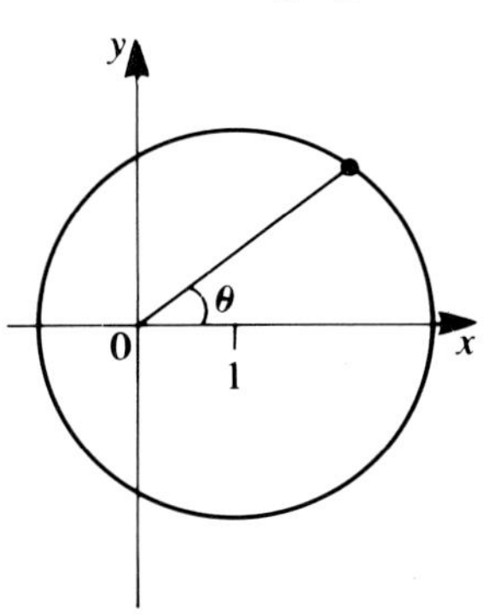

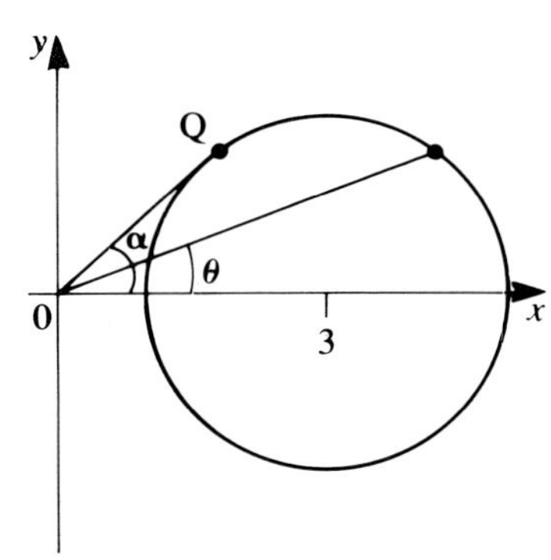

Now $\arg(z)$ is the angle between the x axis and the line joining a point z on the circle to the origin, i.e. angle θ. Clearly in the left figure, θ can take all values, whereas in the right figure, $-\alpha \leqslant \theta \leqslant \alpha$ where OQ is a tangent to the circle.

Problem 77

The equation $|z-2| = 1$ defines the circle of radius 1 and centre (2, 0). Re $(z) > 0$ and Re $(z) \leqslant 3$ for all points in the circle. So strictly speaking Re $(z) < 3$ is not true because Re(z) could equal 3.

Problem 78

The equation $|z-a| = k$ means that the distance between the point representing any complex number z and the point representing the number a is k. As z varies we obtain the set of points on a circle with centre a and radius k.

Problem 79

The equation $\arg(z-a) = k$ means that the angle between the x axis and the line joining the point representing the number z to the point representing a is k. As z varies we obtain the set of points on a straight line through the point representing a, making an angle k with the positive x axis. Note that the half-line obtained does not include the end point, at which $z = a$.

Problem 80

The equation $|z| = 2$ defines the circle shown here.

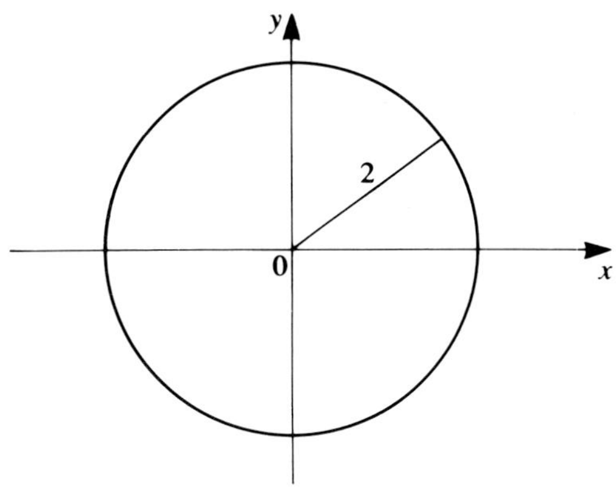

If z is replaced by $\dfrac{1}{z}$ we get a circle of radius $\dfrac{1}{2}$.

Problem 81

$|z| = \sqrt{x^2+y^2}$ is true for all x and y.

$\text{Arg}(z) = \tan^{-1}\left(\frac{y}{x}\right)$ is only true for $x \geqslant 0$.

If $x < 0$ and $y > 0$ then $\arg(z) = \pi - \tan^{-1}\left|\frac{y}{x}\right|$.

If $x < 0$ and $y < 0$ then $\arg(z) = -\left(\pi - \tan^{-1}\left|\frac{y}{x}\right|\right)$.

Problem 82

(a) The point (3, 4) represents the complex number $z = 3 + 4i$. So $iz = -4 + 3i$ which is represented by the point $(-4, 3)$. It is clear that $\hat{AOB} = 90°$ and that multiplying the coordinates by i rotates the line OA through a right angle as shown here.

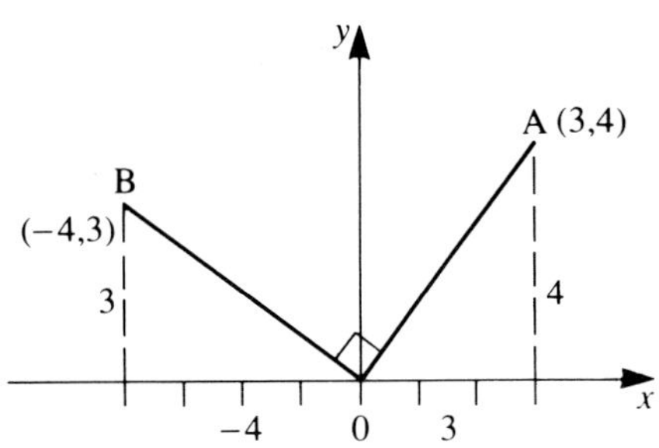

(b) After multiplying by i, the points become, respectively, (0, 0), (0, 2), (−2, 2) and (−2, 0). The square has been rotated about the origin by 90° as the following figure shows.

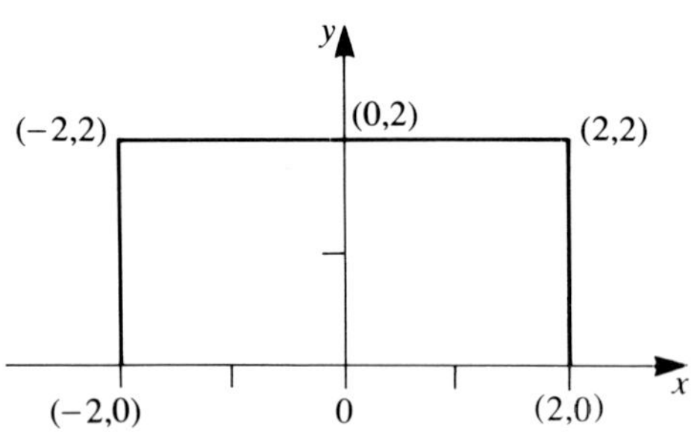

(c) $\frac{1}{i} = \frac{1}{i} \cdot \frac{i}{i} = \frac{i}{i^2} = -i$ and multiplying the coordinates by $\frac{1}{i} = -i$ will rotate the rectangle about the origin by −90°, that is by 90° clockwise (see the figure at the top of the next column).

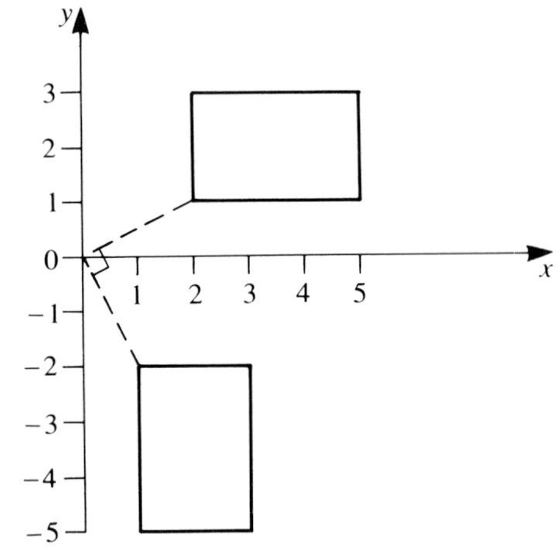

Similar problems
What happens to the following when the coordinates are multiplied by i?

(a) the line from (1, 0) to (3, 5)

(b) the rectangle with vertices (1, 2), (−1, 2), (−1, −2), (1, −2)

(c) the triangle with vertices (1, 1), (5, 1), (1, 4)

Problem 83

The quantity $|z-3|$ measures the distance from the point z to the point (3, 0).

(a) $|z-3| = 2$ means that the distance from the point z to the point (3, 0) is always 2. The locus of z is a circle of radius 2 centred at (3, 0) (see figure below). In cartesians, $z = x + iy$ and hence

$$|(x-3)+iy| = 2$$

$$\text{or} \quad (x-3)^2 + y^2 = 4$$

giving $x^2 + y^2 - 6x + 5 = 0$

which is the cartesian equation of a circle of radius 2 centred at (3, 0).

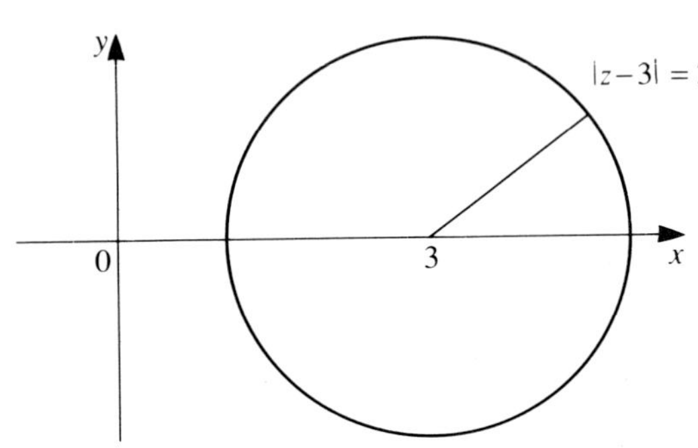

(b) Since $\left|\frac{4}{5}+\frac{3}{5}i\right| = \sqrt{\left(\frac{4}{5}\right)^2+\left(\frac{3}{5}\right)^2} = 1,$

$$\left(z+2i\right) \leqslant \left(\frac{4}{5}+\frac{3}{5}i\right)$$

represents the points inside or on the circle of radius 1 centred at $(0, -2)$ (see below).

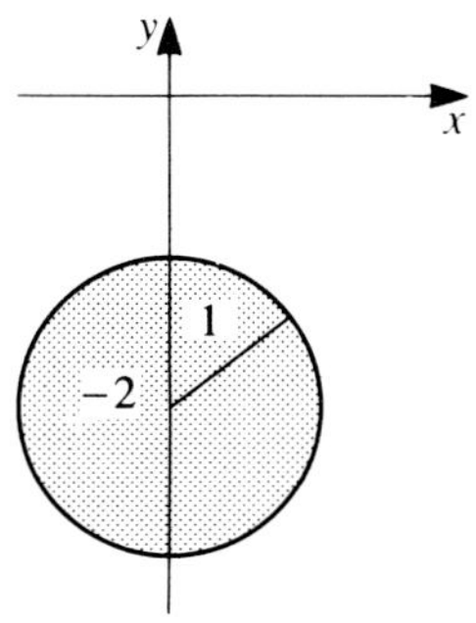

(c) $|z-3| = |z|$ means that the distance from the point z to the point $(3, 0)$ is the same as its distance to the origin. The locus of z is the perpendicular bisector of $(3, 0)$ and $(0, 0)$, which is the line $x = 3/2$.

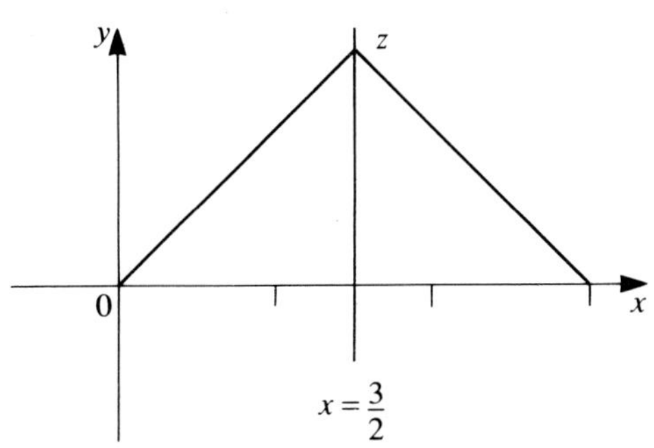

In cartesians, with $z = x + iy$,

$$|(x-3)+iy| = |x+iy|$$
$$(x-3)^2+y^2 = x^2+y^2$$

which simplifies to $x = \frac{3}{2}$.

Similar problems

What is the locus of z if

(a) $|z-i| = e$

(b) $\left|\frac{1}{z}\right| \geqslant 4$

(c) $|z+i| = |z-2|$

Problem 84

The argument of z-i, arg $(z$-$i)$, measures the angle that the line from the point z to the point $(0, 1)$ makes with the positive x axis.

(a) If $\arg(z-i) = \pi/4$ then the locus of z is the straight line through $(0, 1)$ making an angle $\pi/4$ with the positive x axis, as shown here.

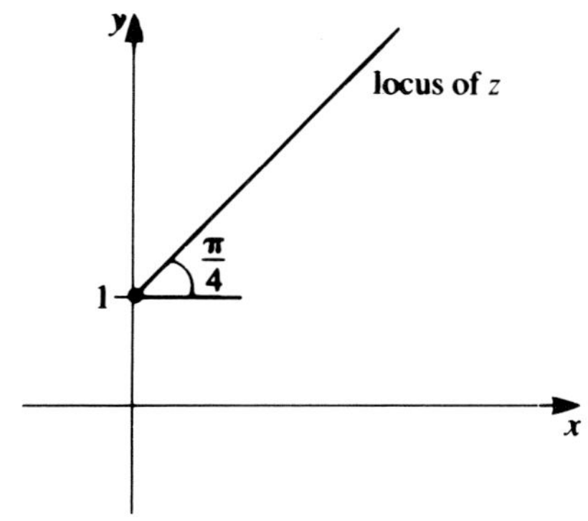

In cartesians, with $z = x + iy$,

$$\arg(z-i) = \arg\left(x+(y-1)i\right)$$
$$= \tan^{-1}\frac{y-1}{x}$$
$$= \frac{\pi}{4} \text{ (given)}$$

Hence $\frac{y-1}{x} = \tan\frac{\pi}{4}$, giving $y = x + 1$.

The locus of z is the line $y = x + 1$ $(x \geqslant 0)$.

(b) If $|\arg(z-1)| \leqslant \frac{\pi}{3}$,

then $-\frac{\pi}{3} \leqslant \arg(z-1) \leqslant \frac{\pi}{3}$.

When $\arg(z-1) = \frac{\pi}{3}$ the locus of z is the straight line from $(1,0)$ making an angle of $\frac{\pi}{3}$ with the positive x axis, as in the figure below.

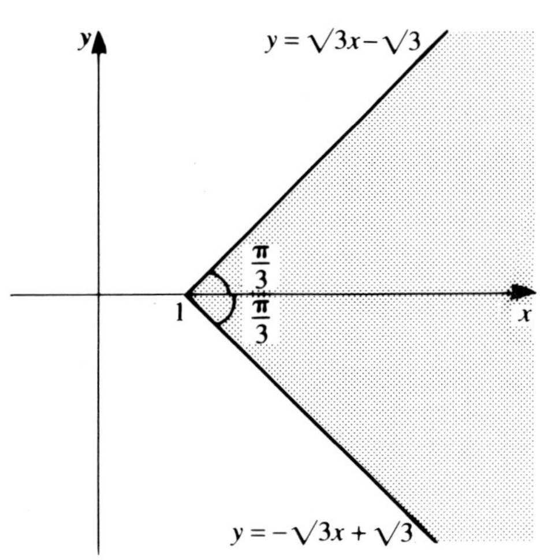

In cartesians,

$$\arg((x-1)+iy) = \frac{\pi}{3}$$

$$\tan^{-1}\frac{y}{x-1} = \frac{\pi}{3}$$

$$\frac{y}{x-1} = \sqrt{3}$$

$$y = \sqrt{3}\,x - \sqrt{3} \quad (x \geqslant 1)$$

Similarly, when $\arg(z-1) = -\frac{\pi}{3}$,

$$y = -\sqrt{3}\,x + \sqrt{3}.$$

Hence $|\arg(z-1)| \leqslant \frac{\pi}{3}$ is as shaded in the diagram.

(c) If $\arg(z+1) = \arg(z+2i)$ then z must lie on the line obtained by extending the line joining $(-1, 0)$ to $(0, -2)$ in either direction.

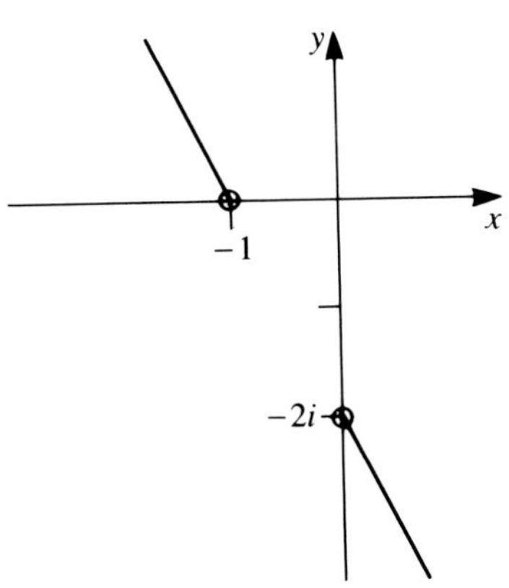

(For points on the line between $(-1, 0)$ and $(0, -2)$, $\arg(z+1) = \pi - \arg(z+2i)$).

In cartesians:

$$\arg((x+1)+iy) = \arg(x+(y+2)i)$$

$$\tan^{-1}\frac{y}{x+1} = \tan^{-1}\frac{y+2}{x}$$

$$\frac{y}{x+1} = \frac{y+2}{x}$$

which simplifies to

$$y = -2x - 2 \qquad (x<-1 \text{ or } x>0).$$

Similar problems

What is the locus of z if:

(a) $|\arg z| \leqslant \frac{\pi}{4}$,

(b) $\arg(z+1) = \frac{\pi}{6}$,

(c) $|\arg z| = |\arg(z-1+2i)|$.

Problem 85

A complex number $z = re^{i\theta}$ has n different nth roots. Their moduli are all $r^{1/n}$, but their arguments,

$\frac{\theta}{n} + \frac{2\pi k}{n}$, $k = 0, 1, \ldots$, differ, one from the next, by $\frac{2\pi}{n}$.

On an Argand diagram the roots are uniformly spaced around the circumference of a circle of radius $r^{1/n}$, each being separated from its neighbour by an angle $\frac{2\pi}{n}$.

Problem 86

No, not every geometric transformation has an inverse – for instance, a shear may not.

The transformation given is not one to one. The points on the square transform into a line with points B′ and D′ being coincident, as shown below.

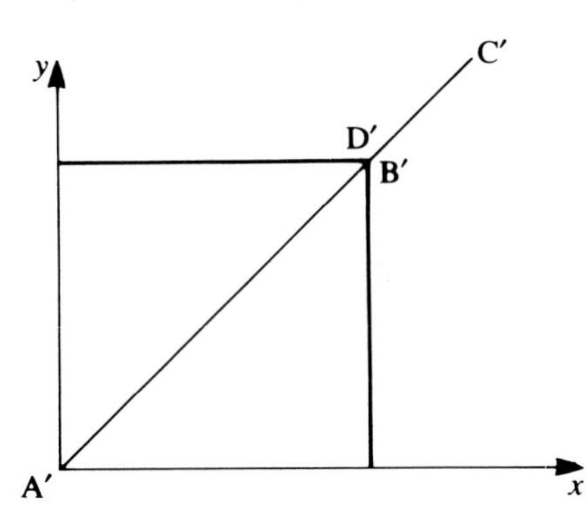

In fact all points on the x and y axes transform onto the line.

Problem 87

(a) QP is a 3×3 matrix.

(b) R is a 2×3 matrix.

(c) $(RQ)^T$ is a 1×2 matrix.

Problem 88

J represents a clockwise quarter-turn.

$$\begin{pmatrix} 1 & 0 \\ 0 & 1 \end{pmatrix} = J^4$$

$$\begin{pmatrix} -1 & 0 \\ 0 & -1 \end{pmatrix} = J^2$$

$$\begin{pmatrix} 0 & -1 \\ 1 & 0 \end{pmatrix} = J^3$$

Problem 89

(i) has determinant a.

$\begin{pmatrix} a & b \\ c & d \end{pmatrix}$ has determinant $ad - bc$.

$\begin{pmatrix} a & b & c \\ a & b & c \\ 0 & 0 & 0 \end{pmatrix}$ has determinant 0.

The other two matrices do not have a determinant because they are not square matrices.

Problem 90

(i), (ii) and (vii) are true provided $|A| = 0$.

If $|A| = 0$ then A^{-1} does not exist.

(iv) and (vi) are sometimes true.

(iii) is always true.

(iv) In general, $(A-B)(A+B) = A^2+AB-BA-B^2$
as $AB \neq BA$ usually.

(v) For example, if $A = \begin{pmatrix} 2 & 2 \\ 1 & 2 \end{pmatrix}$

then $|A| = 2$ and $A^{-1} = \frac{1}{2} \begin{pmatrix} 2 & -2 \\ -1 & 2 \end{pmatrix}$ with $|A^{-1}| = 1$

i.e. $\det(A^{-1}) \neq (\det A)^{-1}$.

This shows that (v) is not always true.

SECTION B

TEACHER MATERIAL

PART 2

Investigations

The following teachers' notes refer to Student Material Part 2,* investigations, *to be found on pages 21 to 36 in this book.

Investigation 1: Tower of Hanoi

Trial and error leads to the following number of moves for 1, 2, 3, 4 discs.

1 3 7 15

The series indicates that the next number might be 31. Checking with 5 discs confirms this. Thus we guess the number of moves for n discs is

$$u_n = 1 + 2 + 2^2 + 2^3 + \ldots + 2^{n-1} = 2^n - 1 \quad \text{(a GP)}$$

So far, the series has only been shown to be true for $n \leqslant 5$.

Note that for 5 discs it takes 7 moves to get to this position:

	1	
4	2	
5	3	–

It takes 15 moves to get to this position:

	1	
	2	
	3	
5	4	–

This is the same as for moving 4 discs.

Then one more move is required to obtain the following:

	1	
	2	
	3	
–	4	5

and a further 15 moves to achieve the objective:

		1
		2
		3
		4
–	–	5

Similarly, for n discs we obtain

$$u_n = u_{n-1} + 1 + u_{n-1}$$

i.e. we obtain the recurrence relation

$$u_n = 2u_{n-1} + 1 \qquad \text{with } u_1 = 1$$

From this and the method of induction, the number of moves can be confirmed as $2^n - 1$,

if $\quad u_n = 2^{n-1} - 1$
then $u_{n+1} = 2u_n + 1 = 2(2^{n-1} - 1) + 1$
$\quad\; u_{n+1} = 2^n - 1.$

Note: Solving this problem practically indicates that there is only one way without going back on oneself. In this sense it is the *smallest* number of moves.

Investigation 2: The cut direct

For a unit cube the following cross-sectional shapes can be made.

Square: cut parallel to a face.

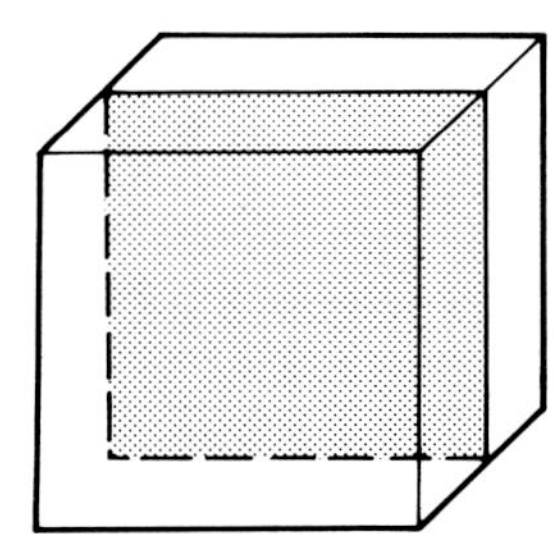

Rectangle: cut perpendicular to a face.

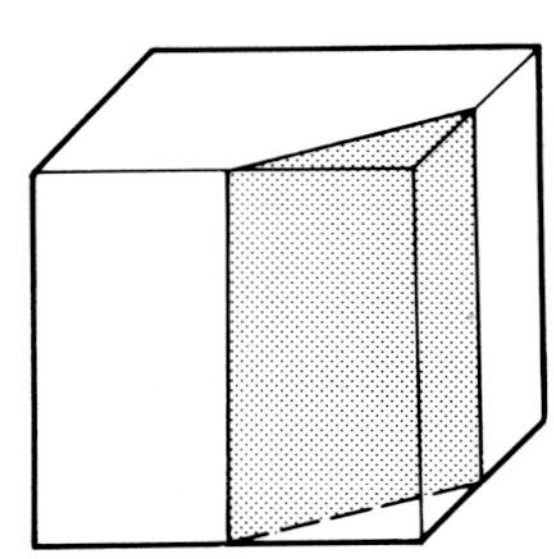

Triangle: cut off a corner.

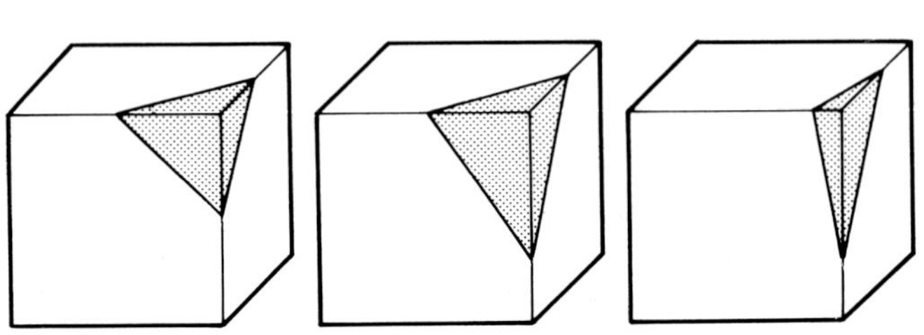

equilateral isosceles scalene

Hexagon: cut past the vertices.

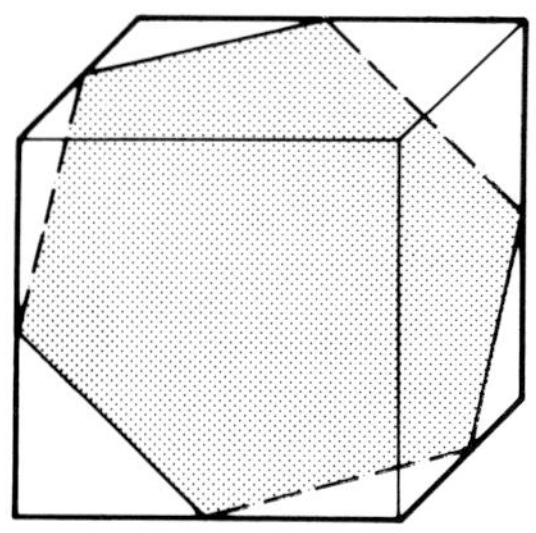

Trapezium: cut past two vertices.

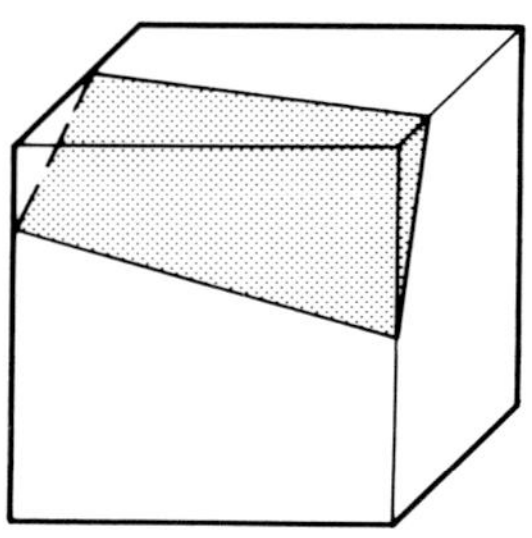

Pentagon: cut past three vertices.

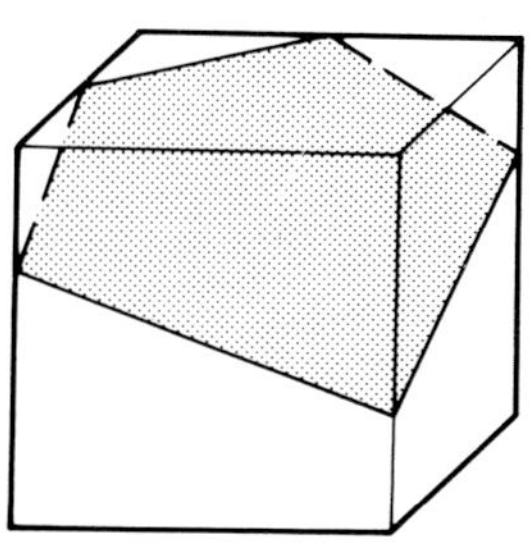

More rectangles: cut from a vertex.

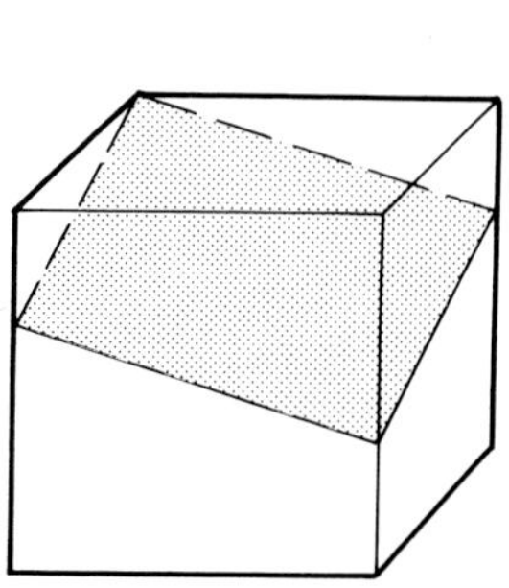

square

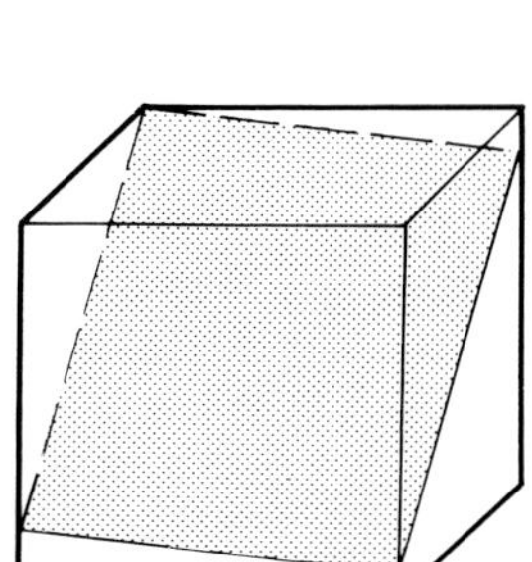

oblong

There may well be others that we have missed. The areas depend on where exactly the cuts are made. The square made by a cut parallel to the face obviously has an area of 1 square unit.

A square made by a cut from a vertex could vary from 1 square unit when $a = 0$ (see the following figure) to $\frac{5}{4}$ square unit when $a = \frac{1}{2}$.

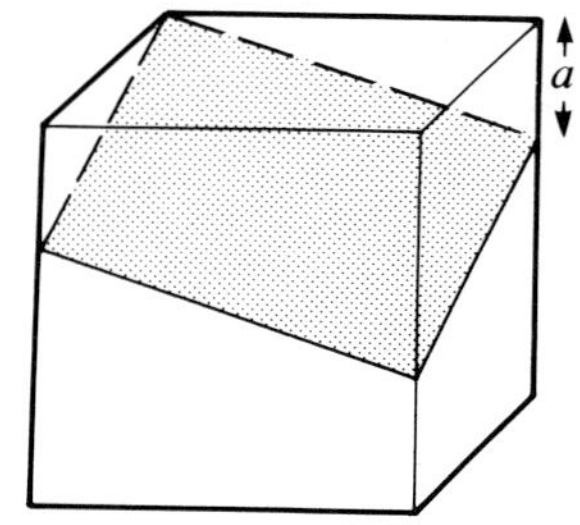

Investigation 3: Cubist painting

2 cm × 2 cm × 2 cm cube

Each cube is painted with three sides of each colour.

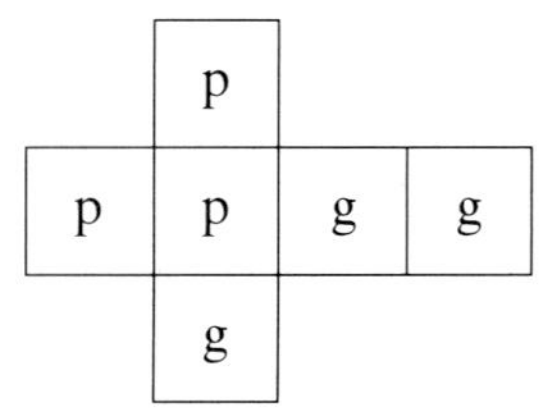

3 cm × 3 cm × 3 cm cube

For each colour you will need the cubes shown in the following figure.

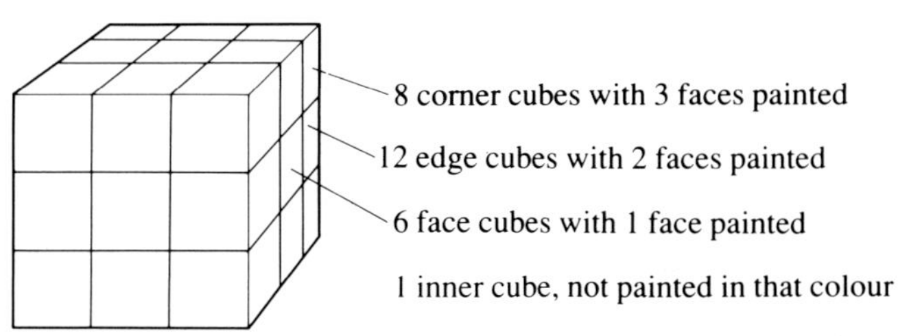

There are various ways of painting the individual cubes and various notations that may be adopted. The easiest way to check a solution in any notation is to check that

(i) each cube has 6 faces painted
(ii) there are 27 cubes
(iii) for each colour there are the cubes as listed above.

Here is one solution (using 'p' for pink, 'g' for green and 'o' for orange).

Net of cube	Number of cubes	Position
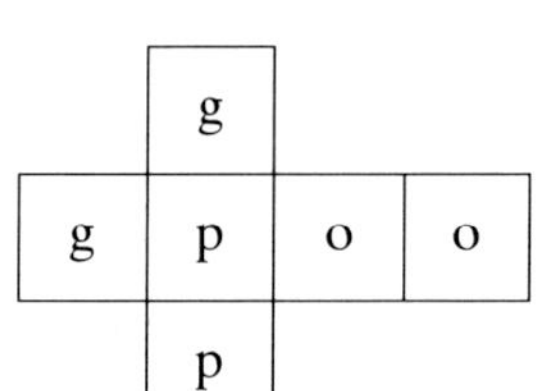	12	Edge cubes for all 3 colourings

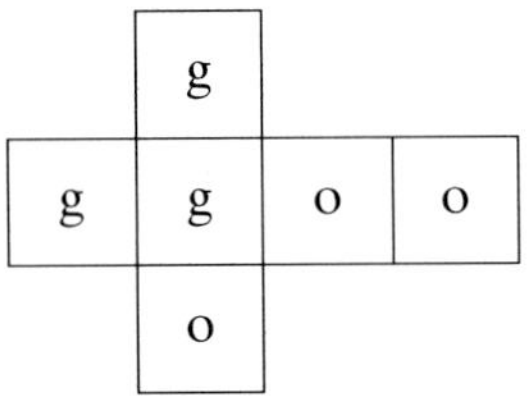

1 Corner cube for green and orange.

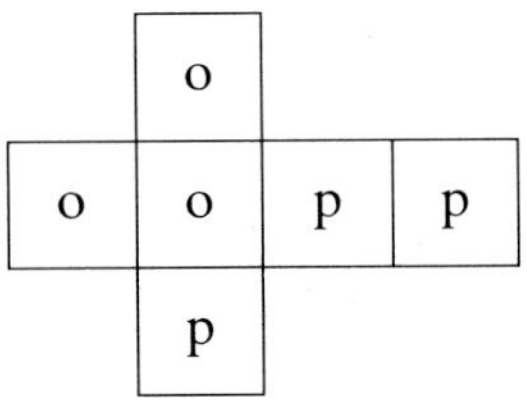

1 Corner cube for orange and pink. Inner cube for green.

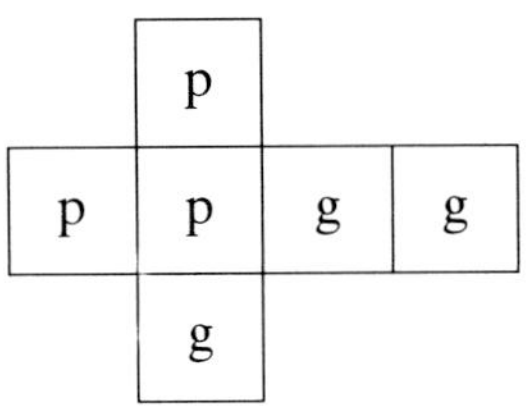

1 Corner cube for pink and green. Inner cube for orange.

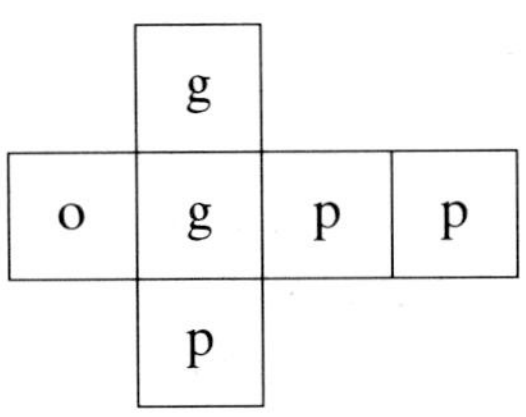

6 Corner cube for pink. Edge cube for green. Face cube for orange.

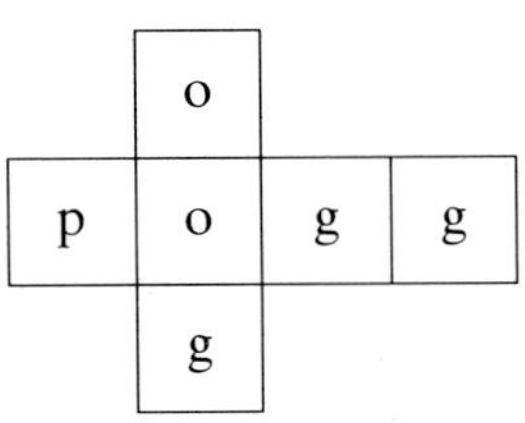

6 Corner cube for green. Edge cube for orange. Face cube for pink.

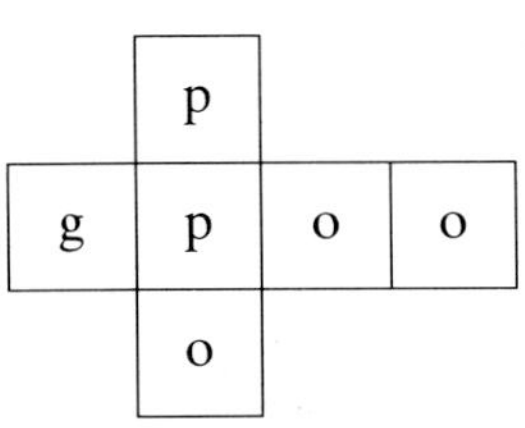

6 Corner cube for orange. Edge cube for pink. Face cube for green.

4 cm × 4 cm × 4 cm cube

This is actually simpler than the 3 cm × 3 cm × 3 cm cube once you have spotted a pattern. Again, the following is *not* a unique solution. For each colour you will need the cubes shown in the diagram at the top of the next column.

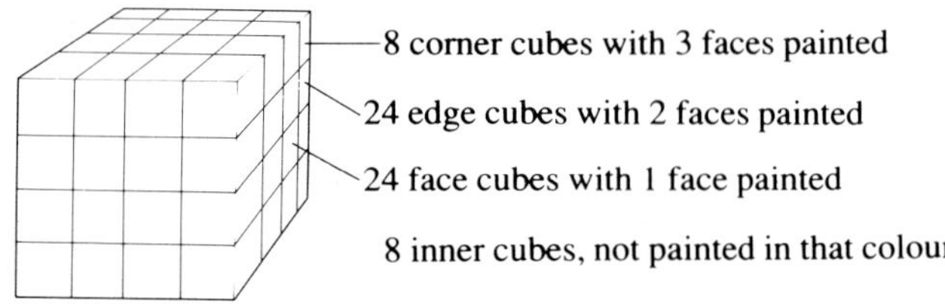

Using 'p' for pink, 'g' for green, 'o' for orange and 'b' for blue:

Net of cube **Number of cubes** **Position**

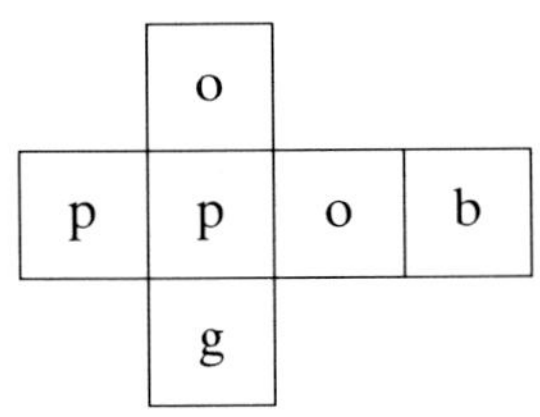

24 Edge cubes for pink and orange. Face cubes for green and blue.

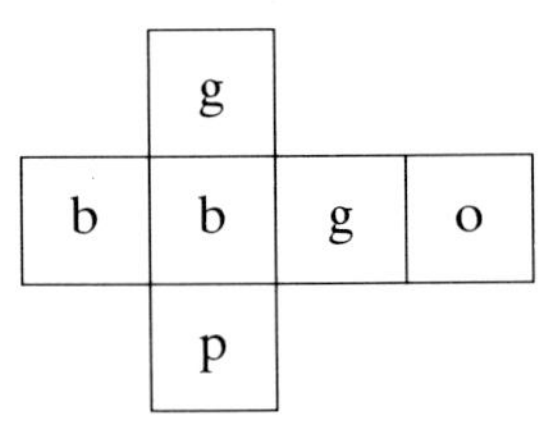

24 Edge cubes for green and blue. Face cubes for pink and orange.

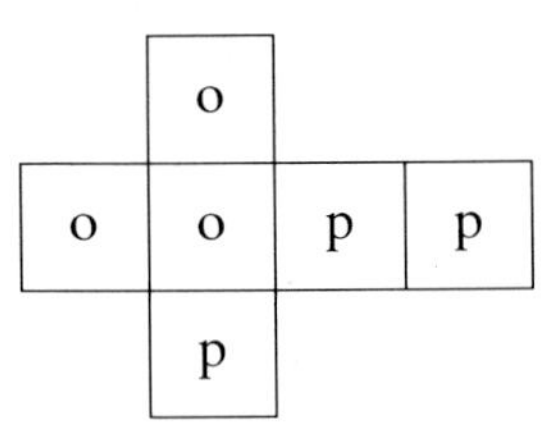

8 Corner cubes for pink and orange. Inner cubes for green and blue.

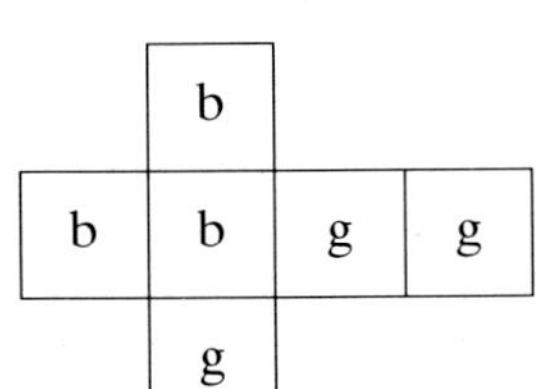

8 Corner cubes for green and blue. Inner cubes for pink and orange.

Investigation 4: Patterns with pins and thread

When $k > n$ there is at least one complete turn round the circle, which we can ignore, plus some step length from 0 to n. For example $n = 7$, $k = 10$ is equivalent to $n = 7$, $k = 3$.

When $k = n$ the thread passes round the outside of all the pins back to the start, as shown below.

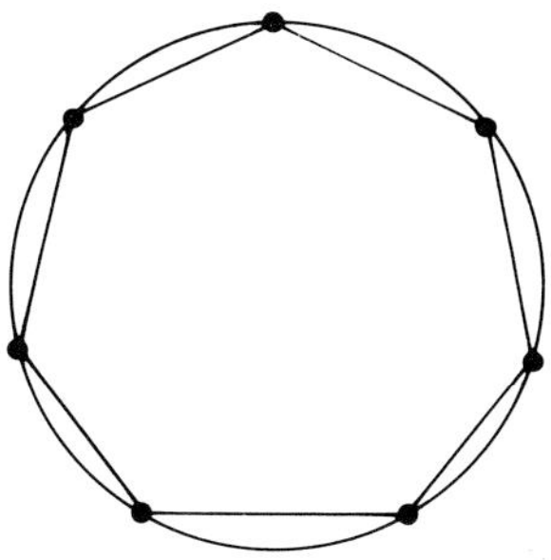

Hence we need only to consider n pins and a step length k, when $k < n$. For example, with $n = 7$ we need only consider $k = 1, 2, 3, 4, 5, 6$.

Once a pin has been reached for the second time a repeat pattern will be obtained and no new pins will be visited. For example, with $n = 6$, $k = 2$ and starting at pin 0, we get 0 2 4 0 2 4 0 . . . and pins 1, 3 and 5 are never reached. This is shown in the figure below.

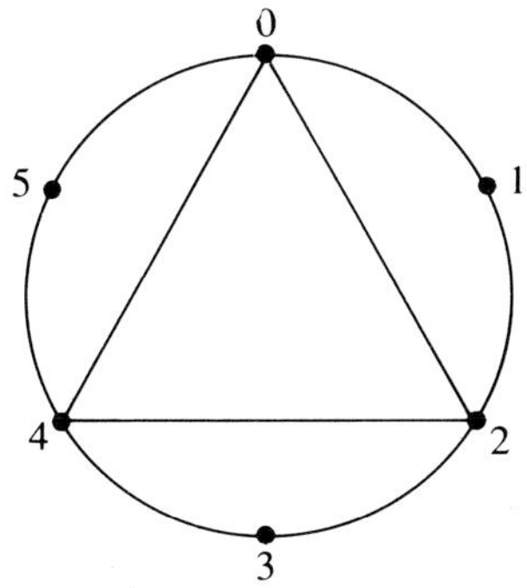

With $n = 7$ then, for all pins to be reached, the first seven pins visited must be some rearrangement of 0, 1, 2, 3, 4, 5, 6. Trying $n = 7$ with $k = 1, 2, 3, 4, 5, 6$ we see that in each case all pins are visited. Trying $n = 6$ it is found that all pins are visited with $k = 1$ or 5, but not with $k = 2, 3$ or 4.

From an investigation of different combinations of n and k it will be seen that not all pins are visited when n and k have a common factor (other than 1). We conjecture that this holds in general; with n pins and a step length k a complete pattern is obtained when, and only when, n and k have no common factor other than 1.

Note: This argument relies on generalising from specific examples. It should be pointed out that the result has not been *proved* correct and that specific examples do not prove general rules. The proof of the result in this investigation involves modular arithmetic and is beyond the scope of an A level course.

Investigation 5: Sharing a pizza

Both methods of cutting the pizza are fairly familiar problems which some students may have investigated as part of their GCSE studies, perhaps as a coursework assignment. However in the sixth form, a more general approach with a higher level of mathematical methods might be expected.

Consider a general approach first which looks at the number of regions formed when the interior of a circle is cut by a family of lines. We develop a notation that will provide a simple rule between the number of regions, the number of lines and the number of intersections.

First consider one line. This produces two regions. Put a small square □ where the line cuts the circle, as shown here.

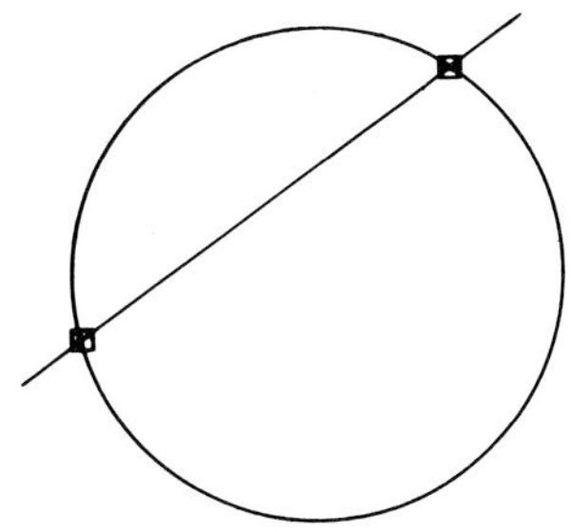

one line

Now if we add a second line, then the number of regions increases by one as the new line cuts the first line and by another one as the line cuts the circle. Put a □ where the line cuts the circle and a small circle ○ at the point of intersection of the two lines, like this.

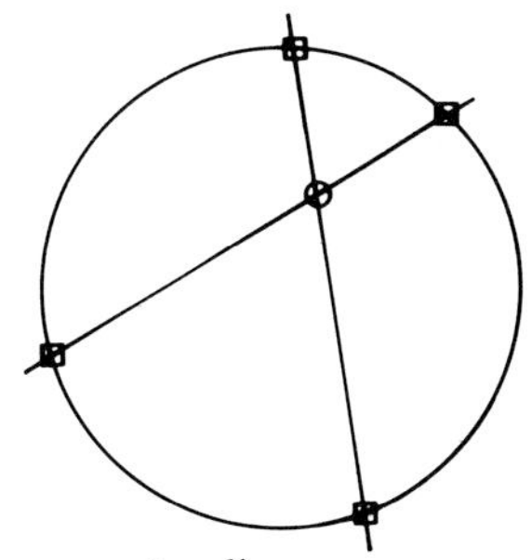

two lines
one intersection

Continue the process.

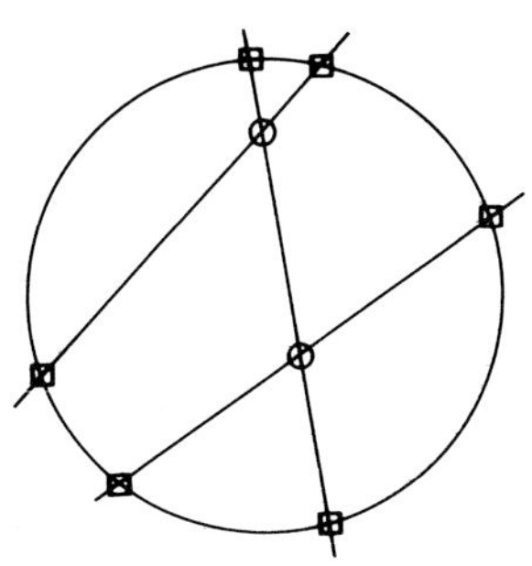

3 lines
2 intersections

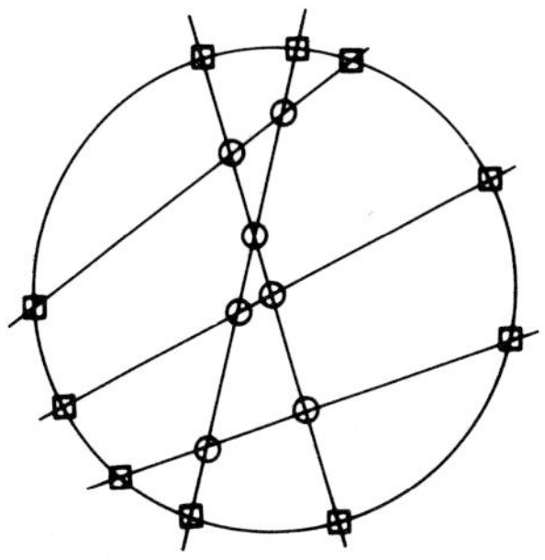

5 lines
7 intersections

Note that the diagrams do not necessarily give the maximum number of regions. As each line is added, the number of regions increases by 1 for each intersection.

We can now find a simple relationship between the number of regions and the number of lines. We have

$$R = 1 + \frac{N}{2} + I$$

where R is the number of regions. N is the number of squares (□) on the diagram and I is the number of intersections (○). (Note that $N/2$ is the number of lines). Clearly, the more intersections there are, the more regions there are.

1 The usual method

Now we can use this formula to solve the first pizza problem. If there are n cuts (i.e. lines) then each cut will produce $(n-1)$ intersections. The maximum number of pizza pieces occurs when none of these intersections are concurrent, as in the figure below.

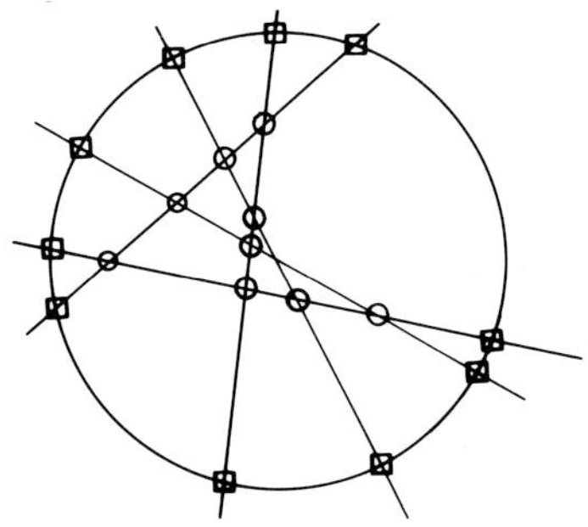

5 lines
10 intersections

Drawing up a table showing the number of intersections and the number of cuts we have:

number of cuts	number of intersections
1	0
2	1
3	3
4	6
5	10
6	15

From this table we can deduce that for n cuts there will be

$$\frac{n(n-1)}{2}$$

intersections. So the maximum number of pizza pieces is given by

$$R = 1 + n + \frac{n(n-1)}{2} = \frac{n^2+n+2}{2}.$$

2 A mathematician's method

This problem is interesting because a pattern spotting method looks hopeful for 1 to 5 points on the circle. We have

number of points	number of regions
1	1
2	2
3	4
4	8
5	16

An unwary student may quickly deduce that $R = 2^{n-1}$. However for 6 points the number of regions is 31 and not 32 and for 7 points there are 57 regions not 64.

Let us try using the general formula

$$R = 1 + \frac{N}{2} + I.$$

The analysis is much harder than for method 1. Consider first the number of lines. Each pair of points produces a line. Take for example 6 points on a circle, as in the figure below.

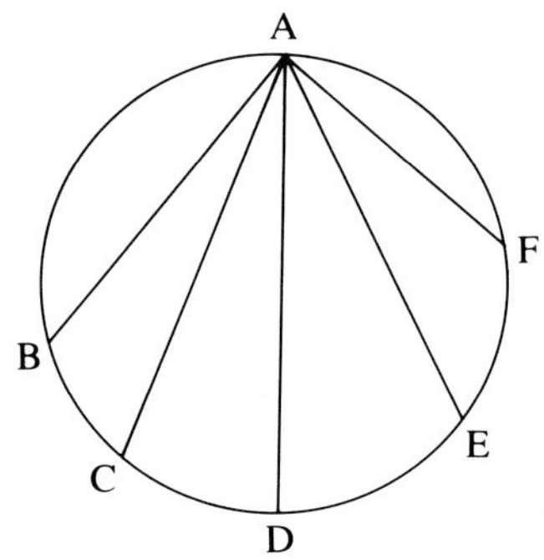

For 6 points, there are 5 lines from each point so the total number of lines is ½ (6×5). In general with n points there will be $n(n-1)/2$ lines.

Now consider the number of intersections. The maximum number of pieces of pizza (i.e. number of regions) will occur when at most two connecting lines intersect (i.e. there are no points of intersection of three or more lines). Each point of intersection is then determined by four points on the edge of the circle, as shown here.

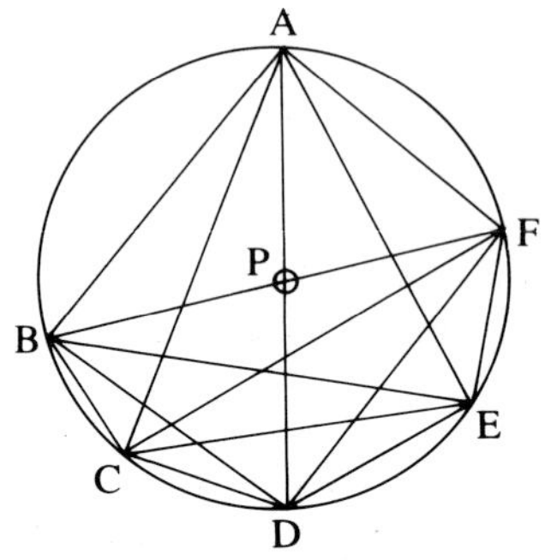

Intersection P is determined by the four points A, C, B, F and joining the 'opposite pairs' of these four points. The number of ways of choosing n things four at a time is given by

$$nC_4 = \frac{n(n-1)(n-2)(n-3)}{24}.$$

If your students have not studied these ideas in their course, then we suggest that they formulate a table showing the number of intersections for various numbers of points, and that they are given the formula to validate. (We cannot imagine many students identifying the rule

$$\frac{n!}{4!\,(n-4)!}$$

without considerable help.) Hence the number of pizza pieces is given by:

$$R = 1 + \frac{n(n-1)}{2} + \frac{n(n-1)(n-2)(n-3)}{24}.$$

Investigation 6: Guess the number

The binary equivalents of the decimal integers are required; for example those for 1, 3, 6 and 11 are 1, 11, 110 and 1011 respectively.

Any integer n in the range 1 to 63 can be expressed as

$$n = a_0.2^0 + a_1.2^1 + a_2.2^2 + a_3.2^3 + a_4.2^4 + a_5.2^5$$

or $\quad n = a_0 + 2a_1 + 4a_2 + 8a_3 + 16a_4 + 32a_5$

where all the a_r are 0 or 1.

Six cards are prepared with 2^r, $r = 0$ to 5 (ie. 1, 2, 4, 8, 16, 32) in the top left hand corner.

A number n will appear on the first card if $a_0 = 1$. Hence the first card contains all odd numbers of the form $2k + 1$.

A number n will appear on the second card if $a_1 = 1$. Hence the second card contains all numbers of the form $4k + 1$ or $4k + 2$.

In general a number will appear on the rth card if $a_r = 1$. For example $26 = 0.2^0 + 1.2^1 + 0.2^2 + 1.2^3 + 1.2^4$ will appear on cards with 2, 8, 16 in the top left hand corner.

Thus when the cards on which a given number appears are known, the number itself can be found by adding together the numbers in the top left hand corners of those cards. The six cards are:

(1)

1	3	5	7	9	11	13	15
17	19	21	23	25	27	29	31
33	35	37	39	41	43	45	47
49	51	53	55	57	59	61	63

(2)

2	3	6	7	10	11	14	15
18	19	22	23	26	27	30	31
34	35	38	39	42	43	46	47
50	51	54	55	58	59	62	63

(3)

4	5	6	7	12	13	14	15
20	21	22	23	28	29	30	31
36	37	38	39	44	45	46	47
52	53	54	55	60	61	62	63

(4)

8	9	10	11	12	13	14	15
24	25	26	27	28	29	30	31
40	41	42	43	44	45	46	47
56	57	58	59	60	61	62	63

(5)

16	17	18	19	20	21	22	23
24	25	26	27	28	29	30	31
48	49	50	51	52	53	54	55
56	57	58	59	60	61	62	63

(6)

32	33	34	35	36	37	38	39
40	41	42	43	44	45	46	47
48	49	50	51	52	53	54	55
56	57	58	59	60	61	62	63

The number 50 appears on cards (2), (5) and (6) and is identified by adding the numbers in the top left hand corner of these cards, $2 + 16 + 32 = 50$.

A similar problem
This is a method of sorting punched cards.

Make a set of 32 (or 16) rectangular cards and punch 5 (or 4) holes along the top of each, as in the figure below. Mark the cards with the numbers 1 to 32 (or 1 to 16).

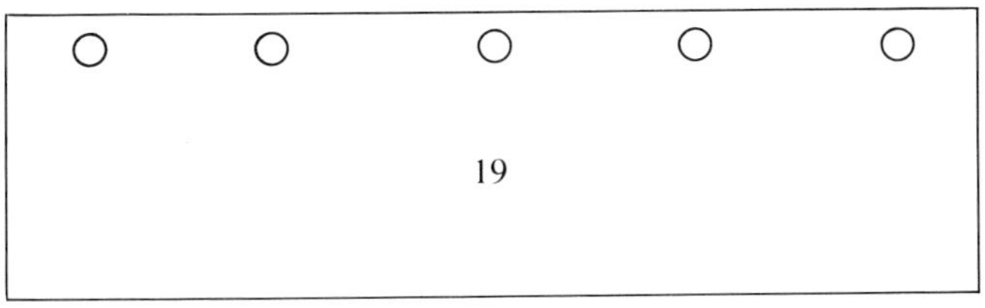

On each card, spaces are then cut above one or more holes (see the figure below), so that the cards can be sorted as explained over the page.

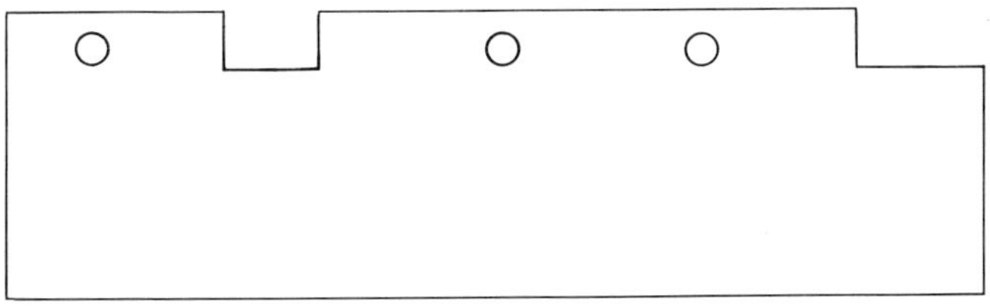

The cards are placed in a pile and a rod (e.g. a pencil) is pushed through the right hand hole. When lifted, this will pull up some of the cards but not others. Those cards that are left are placed behind the others to form a differently ordered pile. This process is repeated, the rod being pushed through the second, third, fourth, fifth hole in turn. At the end of the process the cards should be in numerical order.

The problem is to investigate what combinations of spaces will achieve this.

Investigation 7: The game of NIM

1 Suppose there is one pile with n counters and a player can remove up to 5 counters at a time.

If player A can leave 6 counters then, whatever B does, A can take the rest on his next turn and win. This is a winning position for A. Similarly if A can leave 12 counters then, whatever B does, he can reduce it to 6 on his next turn; hence leaving 12 counters in a winning position for A. It follows that leaving 6, 12, 18, 24, 30, . . . or $6i$, ($i = 1, 2, 3$ . . .) are winning positions.

A's strategy is thus to reduce the pile to a multiple of 6 counters, at which point he or she will be in a winning position. If A goes first he or she can always reach a winning position unless $n = 6i$; in this case he or she should take 1 counter and hope that B does not know the strategy.

Should both A and B know the strategy the winner will depend entirely on the value of n and who goes first:

If $n = 6i$ and A goes first, B wins
If $n \neq 6i$ and A goes first, A wins
If $n = 6i$ and B goes first, A wins
If $n \neq 6i$ and B goes first, B wins

2 Generalising to a player being allowed to take up to k counters at each turn, a winning position is reached when the pile contains $(k+1)\,i$ counters ($i = 1, 2, 3, \ldots$).

3 If the rules are changed so that a player must take at least r but not more than k counters, it follows that the winning positions are when the pile contains $(k+r)i$ counters.

4 With two piles of counters a player can remove counters from one of the piles at his turn. It is easily seen that a winning position is reached when the number of counters in each pile is the same. Then, whatever B does, A just needs to keep the numbers in the two piles equal; as soon as B takes all the remaining counters from one pile, A wins by taking all counters from the other.

5 With three or more piles the situation is more complicated and the complete solution depends on an analysis using binary arithmetic. This is not appropriate for an investigation at this level. Students who develop an interest in the game might like to show that, with three piles of counters, some winning positions are as follows.

1, 2, 3   1, 4, 5   1, 6, 7   1, 8, 9   2, 4, 6   2, 5, 7
2, 8, 10   2, 9, 11   3, 4, 7   3, 5, 6   3, 8, 11   3, 9, 10

They may also find others.

6 For interest and completeness a BASIC program to play NIM is listed below. The user can choose the number of piles of counters, the number in each pile and who is to go first. Unfortunately for him, he is bound to lose, unless he knows the strategy or is very lucky.

```
10  CLS
20  PRINT TAB (30),"*********************"
30  PRINT TAB (30); "*"; TAB (50); "*"
40  PRINT TAB (30); "*"; TAB (38); "NIM"; TAB (50); "*"
50  PRINT TAB (30); "*"; TAB (50); "*"
60  PRINT TAB (30); "*********************"
70  PRINT
80  PRINT TAB (10); "NIM is a game played by two people (you and me) with counters"
90  PRINT TAB (10); "The counters are placed in a number of piles and the players"
100 PRINT TAB (10); "take turns to remove counters from one pile."
110 PRINT
120 PRINT TAB (10); "The winner is the player who removes the last counter"
130 PRINT
140 PRINT TAB (10); "You will be able to decide"
150 PRINT TAB (15); "– the number of piles (from 2 to 10)"
160 PRINT TAB (15); "– the number of counters in each pile (less than 64)"
```

```
170 PRINT TAB (15); "– who is to go first"
180 PRINT
190 PRINT TAB (10); "However, unless you are very clever, I shall win."
200 PRINT
210 PRINT TAB (10); "Press Enter when you are ready to start."
220 INPUT " ", G$
230 CLS
240 INPUT "What is your name"; Z$
250 PRINT
260 INPUT "How many piles of counters would you like";N
270 PRINT
280 If N>1 AND N=<10 AND N=INT (N) THEN 320
290 PRINT "The number of piles must be an integer from 1 to 10; try again"
300 PRINT
310 GOTO 260
320 FOR I=1 TO N
330 PRINT "How many counters do you want in pile"; I;
340 INPUT A(I)
350 PRINT
360 E (I) =A (I)
370 IF A (I) >0 AND A (I) <64 AND INT (A (I)) = A (I) THEN 400
380 PRINT "That is not a valid number of counters; try again": PRINT
390 GOTO 330
400 NEXT I
410 GOSUB 2000
420 PRINT: PRINT: PRINT
430 INPUT "Do you want to go first"; Y$
440 PRINT
450 IF Y$ <> "y" AND Y$ <> "Y" THEN 660
460 PRINT
470 PRINT "**************"
480 PRINT "* YOUR TURN *"
490 PRINT "**************"
500 PRINT
510 INPUT "From which pile would you like to remove counters"; I
520 PRINT
530 IF I <=N AND I>0 AND I=INT (I) THEN 550
540 PRINT I; "is not a valid number of piles; try again": PRINT: GOTO 510
550 IF A (I)=0 THEN PRINT "This pile has no counters; try again": PRINT: GOTO 510
560 PRINT
570 INPUT "How many counters are to be removed"; X
580 IF X <=A (I) AND X>0 AND X= INT (X) THEN 600
590 PRINT X; "is not a valid number of counters; try again"; PRINT: GOTO 550
600 A (I) = A (I) – X
605 E (I) =A (I)
610 Y$="n"
620 GOSUB 2000
630 GOSUB 3000
640 IF M=1 THEN GOSUB 4000 ELSE 450
650 GOTO 260
660 PRINT "************"
670 PRINT "* MY TURN *"
680 PRINT "************"
690 PRINT
700 INPUT "Press Enter when you are ready", G$
710 PRINT
720 FOR I = 1 TO N
730 D (I)=A (I)
```

```
740  FOR J = 5 TO 0 STEP −1
750  IF D (I) < 2 ^J THEN B (I, J)=0 ELSE B (I, J) = 1: D (I) =D (I) − 2 ^J
760  NEXT J, I
770  FOR J=0 TO 5
780  S (J) =0
790  FOR I=1 TO N
800  S (J) = S (J) + B (I, J)
810  NEXT I
820  NEXT J
830  FOR J=5 TO 0 STEP −1
840  F=1
850  IF S (J) = 2*INT (S (J)/2) THEN 980
860  F=2
870  C=J
880  FOR I=1 TO N
890  IF B (I, C)=0 THEN 960
900  B (I, C)=0; R=I: S(C)=1
910  FOR K=C TO 0 STEP −1
920  IF S (K) = 2*INT (S (K)/2) THEN 940
930  IF B (R, K) = 0 THEN B (R,K)=1 ELSE B (R, K)=0
940  NEXT K
950  GOTO 990
960  NEXT I
970  GOTO 990
980  NEXT J
990  IF F=2 THEN 1050
1000 FOR I=1 TO N
1010 IF A (I) = 0 THEN 1040
1020 E (I) =E (I)−1
1030 GOTO 1110
1040 NEXT I
1050 FOR I= 1 TO N
1060 E (I) =0
1070 FOR J=5 TO 0 STEP −1
1080 E (I) =E (I) + 2 ^J*B (I, J)
1090 NEXT J
1100 NEXT I
1110 PRINT TAB (10); "I will take";
1120 FOR I=1 TO N
1130 IF A (I) = E (I) THEN 1170
1140 PRINT TAB (22); A (I) −E (I); "counters from pile"; I
1150 PRINT
1160 A (I) = E (I)
1170 NEXT I
1180 Y$="y"
1190 GOSUB 2000
1200 GOSUB 3000
1210 IF M=1 THEN GOSUB 4000 ELSE 450
1220 GOTO 260
2000 PRINT
2010 PRINT TAB (30); "The state of the game"
2020 PRINT TAB (30); "—————————————"
2030 PRINT
2040 PRINT TAB (25); "Pile        Number of counters"
2050 FOR I = 1 TO N
2060 PRINT TAB (27); I; TAB (44); A (I)
2070 NEXT I
2080 RETURN
3000 M=1
```

```
3010 FOR I =1 TO N
3020 IF A (I)>0 THEN M=2: RETURN
3030 NEXT I
3040 IF Y$="n" THEN 3060
3050 PRINT: PRINT "HARD LUCK – I WIN": V=V+1: GOTO 3070
3060 PRINT: PRINT "CONGRATULATIONS – YOU WIN": W=W+1
3070 PRINT: PRINT
3080 PRINT TAB (30); "Score to date"
3090 PRINT TAB (30); "——————————"
3100 PRINT TAB (25); "I have won"; V; "games"
3110 PRINT
3120 PRINT TAB (25); Z$; "has won"; W; games"
3130 RETURN
4000 PRINT: PRINT
4010 INPUT "Do you want to play again"; P$
4020 PRINT
4030 IF P$="y" OR P$="Y" THEN RETURN
4040 PRINT TAB (20); "********************************"
4050 PRINT TAB (20); "*"; TAB (52); "*"
4060 PRINT TAB (20); "*"; TAB (30); "Final score"; TAB (52); "*"
4070 PRINT TAB (20); "*"; TAB (52); "*"
4080 PRINT TAB (20); "********************************"
4090 PRINT TAB (20); "*"; TAB (52); "*"
4100 PRINT TAB (20); "*"; TAB (30); "I have won"; V; "games"; TAB (52); "*"
4110 PRINT TAB (20); "*"; TAB (52); "*"
4120 PRINT TAB (20); "*"; TAB (30); Z$;" has won"; W; "games"; TAB (52); "2*"
4130 PRINT TAB (20); "*"; TAB (52); "*"
4140 PRINT TAB (20); "********************************"
4150 STOP
```

Investigation 8: Platonic solids

It is helpful to provide some plastic (rigid) regular polygons. The students can find out for themselves what a platonic solid is and they will need to handle them. The hints should be sufficient to get them started. This investigation works well if the students are working together in groups of four or five. It may take an hour or more for the idea to 'click' even with some 'nudges' from you.

For each polygon, each vertex is formed from the incidence of a number of identical regular polygons and all the vertices are identical. By trying out which polygons will form a 'convex' vertex, the students will find how many possible types of vertices there are and hence the number of regular polyhedra (the polygons help to show when some will *not* form a vertex).

Having satisfied themselves practically that there can only be five different vertices, they need to write this up so that someone else can follow the reasoning.

There are two crucial points:

(i) The minimum number of polygons that will form a vertex is 3.

(ii) The sum of the angles of the polygons forming a vertex must be less than 360°. If θ is the angle of the polygon and p the number of polygons forming the vertex, then these conditions may be written as

$$p \geqslant 3$$
$$p\theta < 360°$$

For a triangle, $\theta = 60°$ so p may be 3, 4 or 5.
For a square, $\theta = 90°$ so $p = 4$ only.
For a pentagon, $\theta = 108°$ so $p = 3$ only.
For a hexagon, $\theta = 120°$ so there are no values of p which satisfies the second condition.

Hence there are only 5 regular polyhedra, one with pentagonal faces, one with square faces, and three with triangular faces. Even without the solids in front of them, students will know that three of these polyhedra are the regular tetrahedron and octahedron, and the cube. But from the information above they are unlikely to realise that the remaining two are the dodecahedron and icosahedron. They need to be able to calculate the number of faces.

To continue then, Euler's formula linking the number of vertices (V), the number of edges (E) and the number of faces (F) will be needed

$$V - E + F = 2$$

Two other formulae connecting V with E, V with F or E with F will also be required.

For five triangles forming a vertex:

Each face has 3 edges but each edge belongs to 2 faces so

$$E = \frac{3F}{2}$$

Each face has 3 vertices but each vertex has 5 faces incident to it so

$$V = \frac{3F}{5}$$

This gives $F = 20$, the icosahedron.

For three pentagons at a vertex,

$$E = \frac{5F}{2}$$

$$V = \frac{5F}{3}$$

which gives $F = 12$, the dodecahedron.

To complete this investigation, perhaps the students ought to prove Euler's formula. In the form $N - A + R = 2$ for a connected network, it can be proved by induction on the number of arcs.

Establish that it is true for any given (connected) network. Assume that it is true for a network with k arcs (i.e. $N - k + R = 2$). Then add one more arc. This may be done in two ways:

(i) The arc connects two existing nodes. In this case N remains the same, and R is increased by one so
'$N-A+R$' $= N - (k+1) + (R+1)$
$= N - k + R = 2$.

(ii) The arc drifts into a region ending at a node. Then N is increased by 1, and R remains the same, so '$N-A+R$' $= (N+1) - (k+1) + R$
$= N - k + R = 2$.

Investigation 9: Strange algebra

Following the procedure suggested in Hints and Nudges, the results should be as follows.

(a) $k = 2$

(i) $a = -20$; $x = -1/40$, $y = -3/40$
$a = 19$; $x = -1$, $y = -21/20$
$a = 21$; $x = 1$, $y = 19/20$
$a = 20$; there is no solution.

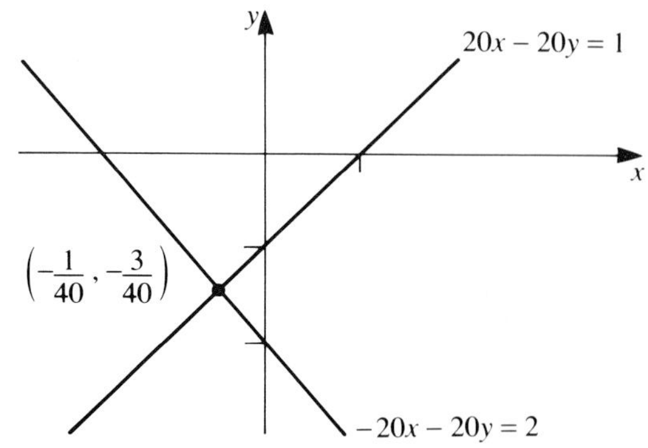

(ii)

In the cases $a = 19$, $a = 21$ the lines are nearly parallel, consequently a small change in the slope of the second equation from $19/20$ to $21/20$ makes a large change in the solution (see the figure below). In the case $a = 20$ the lines are parallel and there is no solution.

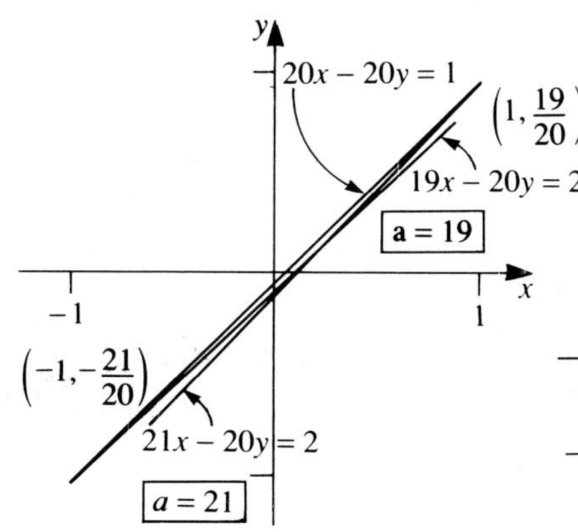

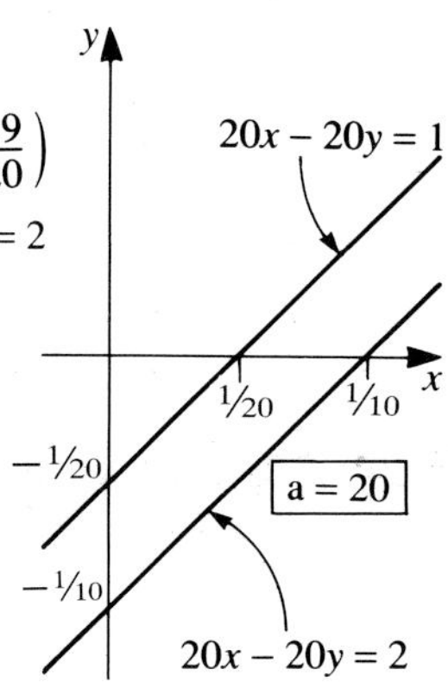

(b) $k = 1$

(i) $a = 20$; $x = \alpha$, $y = \alpha - 1/20$ where α is any number,

$a \neq 20$; $x = 0$, $y = -1/20$

(ii)

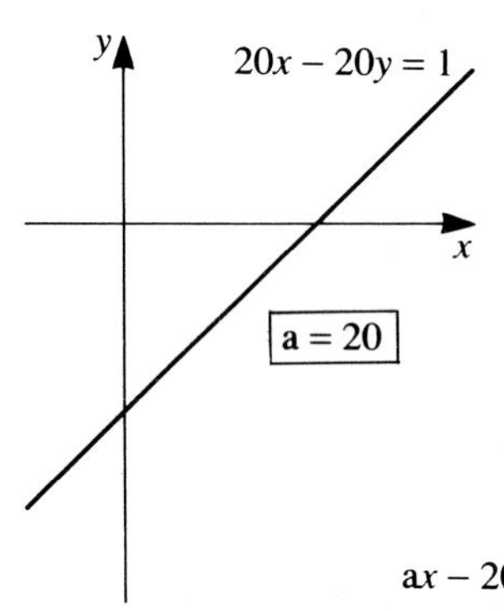

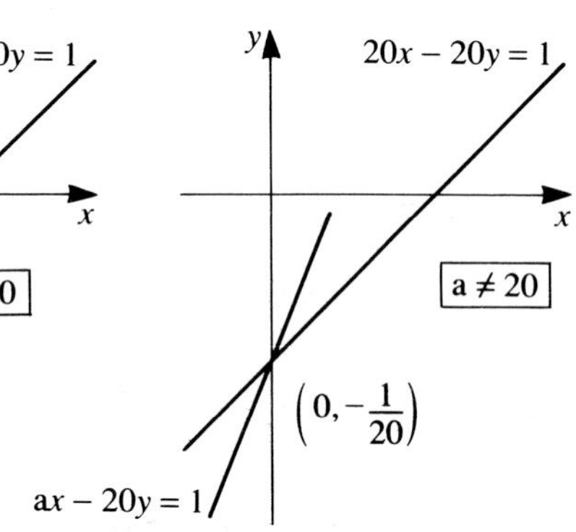

When $a = 20$ the two lines are coincident and there is an infinite number of solutions. When $a \neq 20$ there is a unique solution as shown above.

Investigation 10: Convergence and divergence

Using the approach suggested in Hints and Nudges, the results should be as follows.

1 **(a)** Consider the case $p = ½$. Then a simple BASIC program to find the sum is:

```
10  INPUT N
20  LET S = 0
30  LET T = 1
40  FOR K = 0 TO N
50  LET S = S+ T
60  LET T = T/2
70  NEXT K
80  PRINT S
```

Using this program on a BBC computer gives

the sum to 10 terms is 1.99902344
the sum to 20 terms is 1.99999905

For 30 terms, the computer gives 2, though clearly the accuracy of this result is limited by the computer's internal accuracy.

(b) For $p = 1.1$ we change line 60 in the BASIC program to

```
60 LET T = T * (1.1)
```

The following results are obtained:

$s_{10} = 15.531$
$s_{20} = 64.002$
$s_{30} = 181.943$
$s_{50} = 1281.3$

This is clearly diverging as shown below.

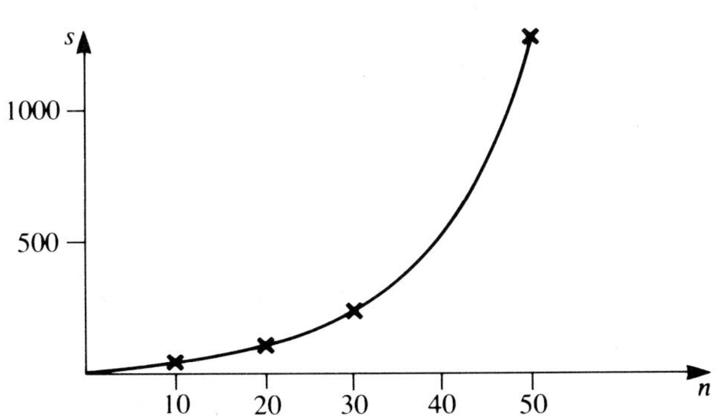

Since the sum is $\dfrac{a(p^n-1)}{p-1} = 10(1.1^n-1)$,

what value of n is required for the sum to be greater than say 10, 100, 1000 . . .?

For $10(1.1^n-1) > 10$ we need

$1.1^n > 2$

$\therefore\ n > \dfrac{\log 2}{\log 1.1}$ i.e. $n > 7.27$ so 8 terms are required.

Similarly for

$10(1.1^n-1) > 100$ $\quad n = 26$

and for $10\,(1.1^n-1) > 1000$ $\quad n = 49.$

Students should be encouraged to explore various positive and negative and fractional values of p. The theoretical result is that the series converges for $|p| < 1$ since

$$\frac{a(1-p^n)}{1-p} \to \frac{a}{1-p} \quad \text{as } n \to \infty \text{ for } |p| < 1.$$

2 **(a)** Consider the case $p = 1$. Line 60 in the program is changed to:

```
40   FOR K = 1 to N
60   LET T = 1/K
```

The following results are obtained:

$s_{10} = 3.829$
$s_{100} = 6.1774$
$s_{1000} = 8.4845$
$s_{10000} = 10.7875$

The series diverges very slowly.

(b) For the series

$$1 + \frac{1}{2} + \left(\frac{1}{4} + \frac{1}{4}\right) + \left(\frac{1}{8} + \frac{1}{8} + \frac{1}{8} + \frac{1}{8}\right) \qquad (1)$$

the sum of the first 4 terms is 2
the sum of the first 8 terms is 2½
the sum of the first 16 terms is 3
the sum of the first 32 terms is 3½.

The series diverges since, by bracketing as shown, the sum of each bracket will be ½ and there is no limit to the number of brackets which can be grouped in this way.

Note: this series cannot be generated from the series in (a). It does however provide a good basic form with which to compare other series (see below).

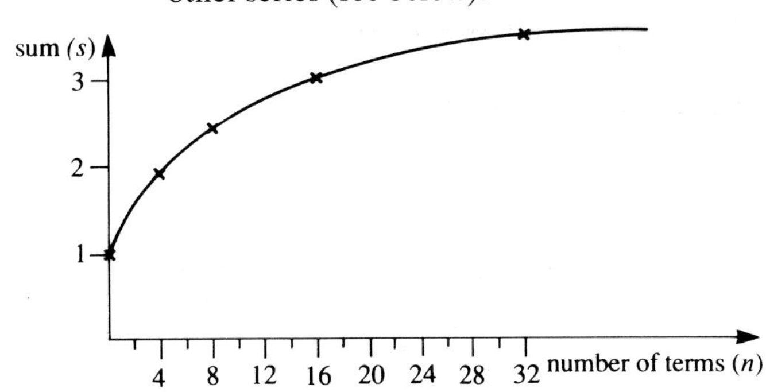

Now let us compare the series

$$s_n = \sum_{k=1}^{n} \frac{1}{k} \text{ with series (1), on the last page.}$$

Grouping the terms in the same way we can write

$$1 + \frac{1}{2} + \left(\frac{1}{3} + \frac{1}{4}\right) + \left(\frac{1}{5} + \frac{1}{6} + \frac{1}{7} + \frac{1}{8}\right)$$
$$+ \left(\frac{1}{9} + \frac{1}{10} + \frac{1}{11} + \frac{1}{12} + \ldots\right)$$

Since $\frac{1}{3} > \frac{1}{4}$, $s_4 > 2$

Since $\frac{1}{5}$, $\frac{1}{6}$, $\frac{1}{7}$ are all greater than $\frac{1}{8}$, $s_8 > 2\frac{1}{2}$

Since the sum of each bracket is greater than the sum of the series (1), the harmonic series must diverge.

For series (1) we have:

n	s_n
1	1
2^1	$1\frac{1}{2}$
$4 = 2^2$	2
$8 = 2^3$	$2\frac{1}{2}$
$16 = 2^4$	3

giving $s_n = 1 + \frac{n}{2}$

So, if $s_n = 10$, $1 + \frac{n}{2} = 10$, $n = 18$.

Therefore if we have 2^{18} terms (i.e. 262144), the sum will exceed 10.

Similarly, if $s_n = 100$, $n = 198$ i.e. we need 2^{198} terms and if $s_n = 1000$, $n = 1998$ we need 2^{1998} terms.

(c) Change the program in **1(a)** as follows:

```
5  INPUT P
30 LET T = 0
60 LET T = K^P
```

Trial and error gives a clear divergence for $p < 0$ and clear convergence for $p > 2$.

We know that if $p = 1$ we have divergence, so it follows that we also have divergence for $p \leqslant 1$ (again using graphical comparison of terms).

For $1 < p < 2$, the picture is less clear. Powerful computers may indicate eventual convergence, but a watertight proof would only be achieved by a more formal treatment using first year undergraduate mathematics (such as the ratio test). This would show that the series does indeed converge in this interval.

Investigation 11: Finding roots

With $a = 2$, the first six terms of the iterative formula are

$x_0 = 1$
$x_1 = 1.5$
$x_2 = 1.41\dot{6}$
$x_3 = 1.414215687$
$x_4 = 1.414213563$
$x_5 = 1.414213563$

If we square x_5 we get $x_5^2 = 2.000000002$. In fact it is easy to show that the iterative formula

$$x_{n+1} = \frac{1}{2}\left(x_n + \frac{2}{x_n}\right)$$

gives an approximation to $\sqrt{2}$ with a wide range of starting values.

Since the formula $x_{n+1} = f(x_n)$ originates from the equation $x = f(x)$, the given formula comes from

$$x = \frac{1}{2}\left(\frac{x+2}{x}\right)$$

which simplifies to $2x^2 = x^2 + 2$
or $x^2 = 2$.

The iterative process can be illustrated using a staircase diagram from which it can be seen that convergence will be rapid for *any* starting value, as shown below.

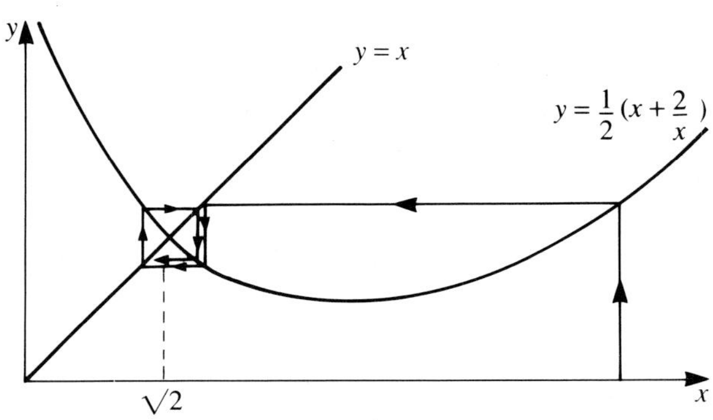

A natural extension to $x^2 = 2$ is $x^2 = a$. Adding x^2 to each side gives

$$2x^2 = x^2 + a$$

which leads to the iterative formula

$$x_{n+1} = \frac{1}{2}\left(x_n + \frac{a}{x_n}\right).$$

Students could be encouraged to find the square root of a using this formula.

Other iterative formulae such as

$$x_{n+1} = \frac{a}{x_n}$$

could be found and explored. (This one does not work very well. For example try $x_0 = 1$).

A further extension is to find the pth root of a number. Starting with

$$x^p = a$$

by adding x^p to each side, we can develop the iterative formula

$$x_{n+1} = \frac{1}{2}\left(x_n + \frac{a}{x_n^{\,p-1}}\right).$$

Formulae of this type were well known in pre-calculator days as quick methods of evaluating roots.

This problem is clearly suitable for a computer investigation. A simple program could be written to generate the solutions of the iterative formulae and to draw the staircase diagrams.

Investigation 12: Chaos!

The mathematics of chaos has become a significant area of study in higher education and has been introduced to the general public through several excellent television programmes. Students can be introduced to the subject at a simple level through the investigation of sequences of the type

$$x_{n+1} = x_n + H\,(x_n - x_n^{\,2})$$

The science of chaos has a wide range of applications, from social science and biological science through to chemical and physical science. The example here is the logistic equation for modelling production growth. The main area of the investigation is to show how a small change in a parameter can lead to strange behaviour in the solution.

A recommended approach to this problem is through a simple computer program or solutions generated on a calculator. The following graph and comments are typical of what can be produced.

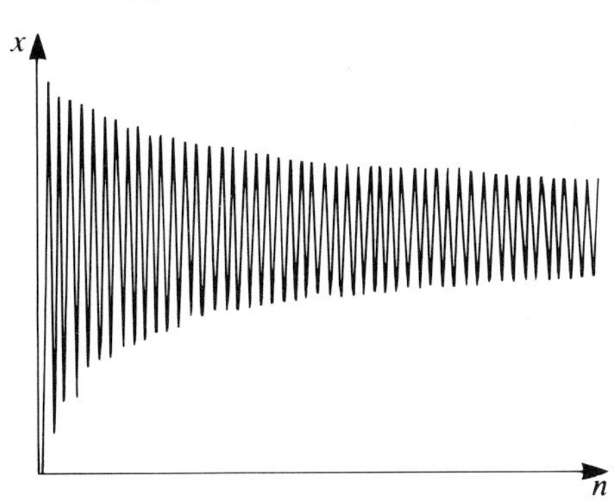

The BASIC program which generated this graph is given below.

```
 10  CLS
 20  PRINT "What is the value of H?"
 30  INPUT H
 40  PRINT "What is the starting value of XO?"
 50  INPUT XO
 60  PRINT H; XO
 70  FOR N=1 TO 1000
 80  XNEW = XO + H* (XO-XO*XO)
 90  PRINT XNEW
100  XO = XNEW
110  NEXT N
120  END
```

The investigation requires students to explore different values of H and x_0.

$H = 2$ is the critical value in this iterative formula. If $H < 2$ the iterative formula is said to have a stable point $x = 1$. To illustrate the idea of a stable point consider the graphical solution of the equation

$$x = x + Hx\,(1-x)$$

The figure below shows the graphs of $y = x$ and $y = x + Hx\,(1-x)$ for $H = 1.5$. Where the graphs intersect is called a **fixed point**. There are two fixed points in this example, $x = 0$ and $x = 1$.

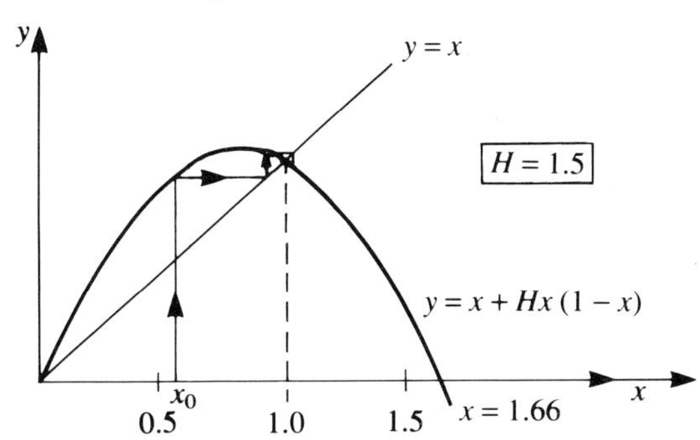

To determine if the fixed point is stable, we draw a vertical line from some point on the x axis (this is x_0 in the iterative formula) until the line intersects the curve. Then we draw a horizontal line until we meet $y = x$. Next we draw a vertical line until it meets the curve, and so on. The fixed point at $x = 1$ is stable because the **cobweb diagram** we have constructed has converged on one point only.

Now if $H > 2$ the situation is different. The figure below shows the cobweb diagram for $H = 2.5$.

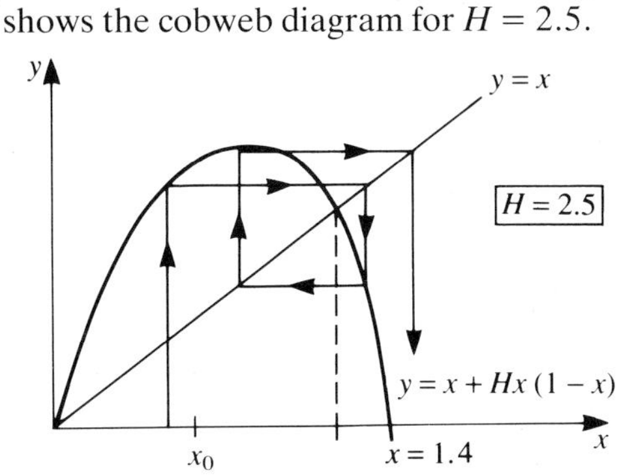

The construction in this case does not converge on the fixed point $x = 1$. This is an example of an unstable fixed point.

The stability of a fixed point depends on the gradient of the function at that point, in this case the gradient of $y = x + Hx\,(1-x)$ at $x = 1$. Generally if the iterative formula is $x_{n+1} = f(x_n)$ then a fixed point x^* is a solution of the equation $x = f(x)$.

If $\left|\dfrac{df}{dx}\right| < 1$ at x^* then x^* is a **stable** point.

If $\left|\dfrac{df}{dx}\right| > 1$ at x^* then x^* is an **unstable** point.

If $\left|\dfrac{df}{dx}\right| = 1$ at x^* then x^* is said to be **metastable**.

For the iterative scheme in this investigation

$f(x) = x + Hx\,(1-x)$ and $x^* = 1$

$$\frac{df}{dx} = 1 + H(1-2x)$$

Now at $x = 1$, $\left|\dfrac{df}{dx}\right| = |1-H| < 1$ (or > 1) if $H < 2$ (> 2).

Thus we can see why $H = 2$ is a critical value for the solution of the iterative formula. If $H < 2$, the formula converges to $x_n = 1$ for any starting value x_0. For $H > 2$ the formula diverges.

Investigation 13: Trapezoidal errors grow!

Integration by parts gives

$$I_r = \left[-x^r e^{-x}\right]_0^1 + r\int_0^1 x^{r-1}e^{-x}\,dx$$

Hence $I_r = rI_{r-1} - \dfrac{1}{e}$.

Using the recurrence relation again

$$I_r = r\left((r-1)I_{r-2} - \frac{1}{e}\right) - \frac{1}{e}$$

$$= r(r-1)\left((r-2)I_{r-3} - \frac{1}{e}\right) - \frac{1}{e}\,(1+r)$$

i.e. $I_r = r!\,I_0 - \dfrac{1}{e}\,(1 + r + r(r-1) + \ldots + r!)$ (1)

where $I_0 = 1 - \dfrac{1}{e}$.

With $r = 1, 2$ and 3 we obtain the following values for I_1, I_2 and I_3.

$$I_1 = 1 - \frac{2}{e},\; I_2 = 2 - \frac{5}{e} \text{ and } I_3 = 6 - \frac{16}{e}$$

(also $I_4 = 24 - \dfrac{65}{e}$, $I_5 = 120 - \dfrac{326}{e}$, $I_6 = 720 - \dfrac{1957}{e}$ — these are not for the faint-hearted!)

Consider the case $r = 6$. The following table gives values of $I_6 \times 10^2$ generated by computer using the trapezoidal rule, with 10, 100, 1000 and 10 000 intervals together with the value of $I_6 \times 10^2$ calculated from (1).

e \ n	10	100	1000	10000	I_6 calculated from (1)
2.7	6.18261	6.02987	6.02834	6.02833	−481.481
2.72	6.14303	5.99163	5.99012	5.99010	51.471
2.718	6.14696	5,99543	5.99391	5.99390	−1.47165
2.7183	6.14637	5.99486	5.99334	5.99333	6.47465
2.71828	6.14641	5.99490	5.99338	5.99337	5.94495
full accuracy of computer	6.14641	5.99490	5.99338	5.99336	5.99360

The exact result is 5.99336. The inaccuracy when using (1) arises because it is necessary to subtract two nearly equal numbers.

Equation (1) can instead be used as follows:

$$I_r = r!\left(1 - \frac{1}{e}\right) - \frac{1}{e}\,(1 + r + r\,(r-1) + \ldots + r!)$$

$$= r! - \frac{1}{e}\,(1 + r + r(r-1) + \ldots + 2r!)$$

$$= r! - r!\left(\frac{1}{r!} + \frac{1}{(r-1)!} + \frac{1}{(r-2)!} + \ldots + \frac{1}{2!} + 2\right)\frac{1}{e}$$

$$= r! - r!\left(e - \left(\frac{1}{(r+1)!} + \frac{1}{(r+2)!} + \frac{1}{(r+3)!} + \ldots\right)\right) - \frac{1}{e}$$

$$= \frac{1}{e}\left(\frac{1}{r+1} + \frac{1}{(r+2)(r+1)} + \frac{1}{(r+3)(r+2)(r+1)} + \ldots\right)$$

Using the first seven terms in the case $r = 6$ we now obtain a more accurate result:

$I_6 = 5.99336$.

Investigation 14: Rough cuts

Since the angle at P is a right-angle we need to consider all positions of P on the semicircle shown in the figure below.

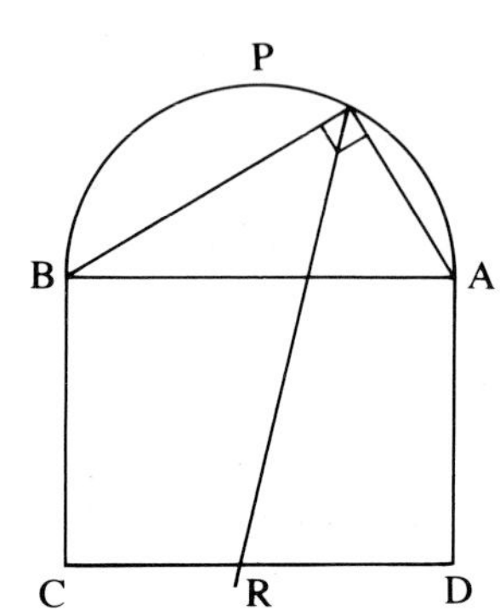

When AP = PB, the bisector of angle P bisects AB and CD so clearly PR divides the square into two equal areas.

As P→ A, angle PAB → 90° and the bisector PR becomes more and more parallel to the diagonal AC i.e. as P → A, R → C. In the limiting case PR again cuts the square in half. A similar argument applies as P → B since R → D. Alternatively we can prove that if R is at C then angle PBA = 0, as shown here.

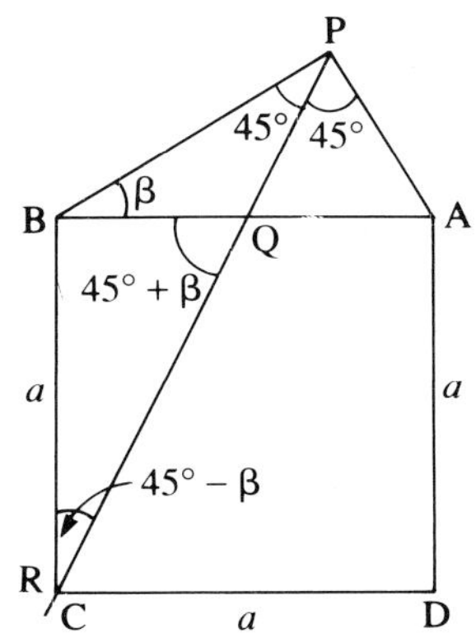

$$BQ = \frac{a}{\tan(45°+\beta)} \quad \text{(from } \Delta\,CBQ)$$

$$BP = a\cos\beta \qquad \text{(from } \Delta\,ABP)$$

Applying the sine rule to Δ BPQ gives

$$\frac{BQ}{\sin 45°} = \frac{BP}{\sin(45°+\beta)}$$

$$\frac{\sqrt{2}\;a}{\tan(45°+\beta)} = \frac{a\cos\beta}{\sin(45°+\beta)}$$

Hence $\sin(45°+\beta) = 0$ or $\sqrt{2}\cos(45°+\beta) = \cos\beta$.

Since $0 \leqslant \beta \leqslant 90°$, $\sin(45°+\beta) \neq 0$. Thus

$$\sqrt{2}\cos(45°+\beta) = \cos\beta$$
$$\cos\beta - \sin\beta = \cos\beta$$
i.e. $$\sin\beta = 0.$$

So β = 0 if R is at corner C. It is therefore evident that R must lie on CD.

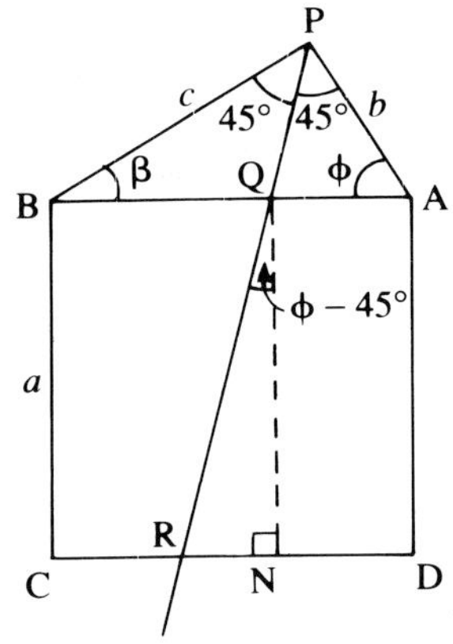

We now require the position of R. Consider triangle RNQ,

$$RN = a\tan(\varphi-45) = \frac{a(c-b)}{c+b}$$

Applying the sine rule to ΔPAQ gives

$$AQ\sqrt{2} = \frac{b}{\sin(135-\varphi)}$$

$$AQ = \frac{ab}{b+c} \qquad (1)$$

Hence $RD = RN + ND = RN + AQ = \dfrac{ac}{b+c}$.

Applying the sine rule to ΔBQP (or by analogy with (1)):

$$BQ = \frac{ac}{b+c} = RD.$$

Thus PQR bisects the area of ABCD, and divides triangle BPA in the ratio AQ: BQ = b:c.

Investigation 15: Integer triangle areas

The method of trial and error will quickly result in the case with sides 3, 4, 5 units. This method is not practical for finding other cases.

A second approach can be made by considering two right-angled triangles together to form a more general triangle, as in the figure below. In this case $(d + b)$, c and e must be consecutive integers such that at least one of a and $(d+ b)$ is even. Then using (½ × base × perpendicular height) gives the required result. In fact, from the figure below, it can be shown that $(d+b)$ is always even.

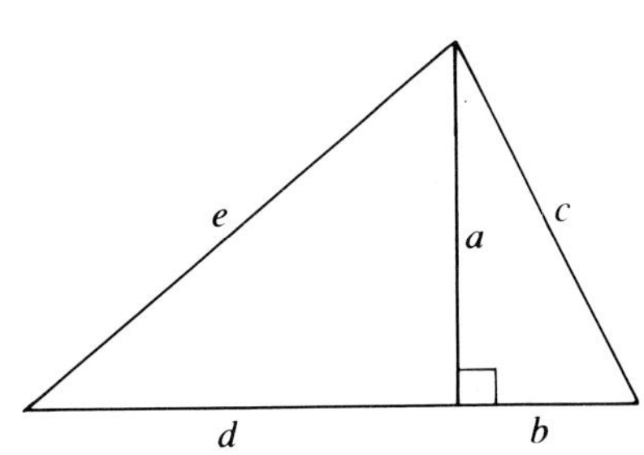

Here (a, b, c) and (a, d, e) are pythagorean triples. This method is very demanding but has its rewards.

A third approach using Heron's formula and the notation in the figure overleaf gives:

$$\text{area} = \tfrac{1}{4}n\sqrt{3(n^2-4)}$$

from which we can show that n will have to be even for $3(n^2-4)$ to be a square.

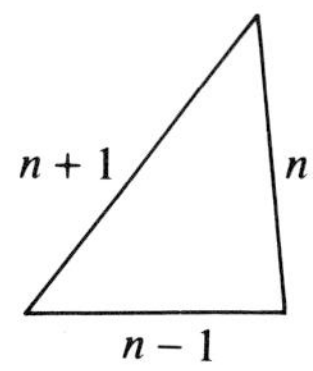

Other cases follow from this by trial and error of values of n but further reduction of cases can take place.

Investigation 16: Loci

For this investigation, students will need several copies of the sheet of circles enclosed with this pack. All of this work can be done as loci on the Cartesian plane by interpreting $|z-a|$ as the length of the line AP.

Students need to know that $|z-a|$ gives the distance between the points representing z and a if the Argand diagram is being used.

All of the loci produced are conics: circles, ellipses and hyperbolae (assuming that the perpendicular bisector of AB can be seen as an infinite circle). It is possible for students to find the Cartesian equations of the curves they draw without knowing the traditional equations for the conic sections, but this is a far more satisfying exercise if students can recognise the shapes of the equations found.

The solutions that follow correspond to the suggestions in part 2 of the Hints and Nudges for this investigation.

(a) $|z-a| = k$
AP $= k$

This defines a circle, centre $z = a$ and radius k. k must be positive.

Suppose that a is the point (a_1, a_2) then

$$|z-a| = \sqrt{(x-a_1)^2+(y-a_2)^2} = k$$

Thus the equation of the locus is
$x^2 + y^2 - 2a_1x - 2a_2y + (a_1{}^2+a_2{}^2-k^2) = 0$

(b) $|z-a| = |z-b|$
AP = BP

For AP and BP to be equal, P must lie on the perpendicular bisector of AB.

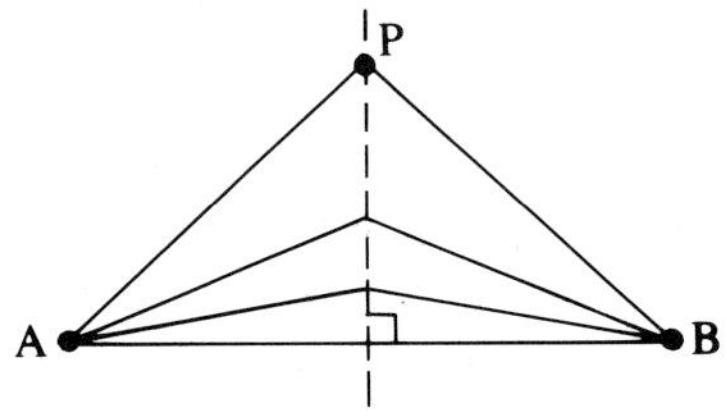

Suppose that A and B are the points (a_1, a_2) and (b_1, b_2) respectively; then:

$$AP^2 = (x-a_1)^2 + (y-a_2)^2$$
and $$PB^2 = (x-b_1)^2 + (y-b_2)^2.$$

Equating AP^2 and PB^2 we have

$$(x-a_1)^2 + (y-a_2)^2 = (x-b_1)^2 + (y-b_2)^2.$$

Expanding the brackets and simplifying gives

$$2(a_1-b_1)x + 2(a_2-b_2)y = a_1{}^2 + a_2{}^2 - b_1{}^2 - b_2{}^2.$$

This is the equation of a straight line.

(c) $|z-a| = k|z-b|$
AP $= k$BP

k must be positive. If $k = 1$ we have case **(b)**.
If $k = 1$ we obtain a circle. If $k > 1$ then B is inside the circle. If $0 < k < 1$ then A is inside the circle.

Suppose that A and B are the points (a_1, a_2) and (b_1, b_2) respectively, then AP $= k$BP so that

$$(x-a_1)^2 + (y-a_2)^2 = k^2[(x-b_1)^2 + (y-b_2)^2].$$

Expanding the brackets and simplifying gives

$$(k^2-1)x^2 + (k^2-1)y^2 + 2x(a_1-k^2b_1)$$
$$+ 2y(a_2-k^2b_2) + k^2(b_1{}^2+b_2{}^2) - (a_1{}^2+a_2{}^2) = 0.$$

Writing this in standard form for the equation of a circle we have

$$\left(x + \frac{a_1-k^2b_1}{k^2-1}\right)^2 + \left(y + \frac{a_2-k^2b_2}{k^2-1}\right)^2 =$$

$$\frac{k^2}{(k^2-1)^2}[(a_1-b_1)^2+ (a_2-b_2)^2].$$

This is the equation of a circle, centre

$$\left(\frac{k^2b_1-a_1}{k^2-1}, \frac{k^2b_2-a_2}{k^2-1}\right)$$

and radius R

where $$R^2 = \frac{k^2}{(k^2-1)^2}[(a_1-b_1)^2+(a_2-b_2)^2].$$

Now $(a_1-b_1)^2 + (a_2-b_2)^2 = AB^2$ so that the radius of the circle is

$$\left|\frac{k}{(k^2-1)}AB\right| .$$

The students should be encouraged to explore the geometrical significance of these results.

As an example, suppose that the origin is chosen as the midpoint of AB, and let A and B lie on the line through the origin such that $a_1 = -c$, $a_2 = 0$, $b_1 = c$ and $b_2 = 0$.

Then if $k < 1$, the centre of the circle lies on the line BA produced, at the point

$$\frac{-c(1+k^2)}{(1-k^2)}, 0$$

and the radius is given by

$$R = \frac{2ck}{(1-k^2)}.$$

We can show that $AC = kR$ and $BC = \frac{R}{k}$, so the circle contains the point A.

Similarly, if $k > 1$ then the circle lies on the line AB produced, and the circle contains the point B.

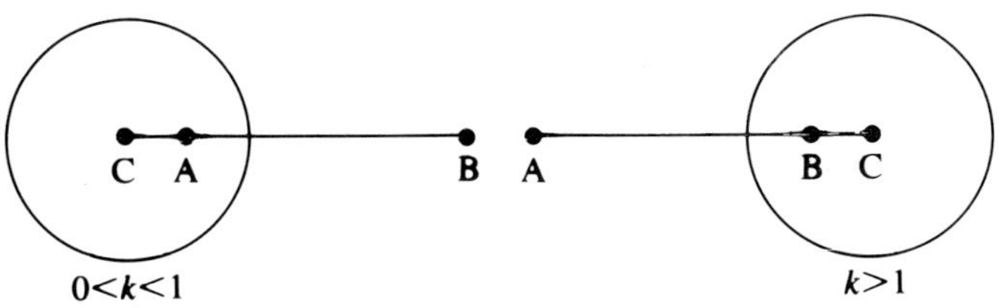

(d) $|z-a| + |z-b| = k$
$AP + BP = k$

If k is larger than the distance between A and B, this is an ellipse with foci at A and B.

For general coordinates A and B, the algebra is rather lengthy; instead suppose that we choose A and B as the points $(-a,0)$ and $(a,0)$ respectively. Then

$$|z-a| = \sqrt{(x+a)^2+y^2}$$

and $|z-b| = \sqrt{(x-a)^2+y^2}$

so that $|z-a| + |z-b| = k$ becomes

$$\sqrt{(x+a)^2+y^2} + \sqrt{(x-a)^2+y^2} = k.$$

Squaring and simplifying gives

$$(x^2+a^2+y^2) + \sqrt{(x+a)^2+y^2}\,\sqrt{(x-a)^2+y^2} = \frac{k^2}{2}. \qquad (1)$$

Rearranging this equation and squaring again we have

$$[(x+a)^2+y^2][(x-a)^2+y^2] = \left[\frac{k^2}{2} - (x^2+y^2+a^2)\right]^2$$

$$(x^2-a^2)^2 + y^2[(x-a)^2 + (x+a)^2] + y^4 = \frac{k^4}{4} - k^2(x^2+y^2+a^2) + (x^2+y^2+a^2)^2.$$

Expanding the brackets and simplifying this reduces to

$$(k^2-4a^2)x^2 + k^2y^2 = \frac{k^4}{4} - k^2a^2$$

which can be written as

$$\frac{x^2}{\alpha^2} + \frac{y^2}{\beta^2} = 1$$

where $\alpha^2 = \frac{k^2}{4}$ and $\beta^2 = \frac{k^2}{4} - a^2$.

This is the equation of an ellipse with foci at A and B. The lengths of the axes are k and $\sqrt{k^2-4a^2}$ provided $k > 2a$. Hence k must be greater than the distance between the two points A and B.

(e) $|z-a| - |z-b| = k$
$AP - BP = k$

This is an ellipse or a hyperbola with foci A and B.

The analysis is similar to the case **(d)** above giving

$$\frac{x^2}{\alpha^2} + \frac{y^2}{\beta^2} = 1$$

where $\alpha^2 = \frac{k^2}{4}$ and $\beta^2 = \frac{k^2}{4} - a^2$.

Investigation 17: Snowflakes

In order to determine the length of the snowflake curve and the area it would enclose, it is necessary to determine

(i) the number of straight line segments

(ii) the length of each straight line segment

(iii) the area of each new equilateral triangle added to the snowflake at each stage.

After some experimentation it becomes evident that

(i) each straight line segment generates 4 new segments at each stage.

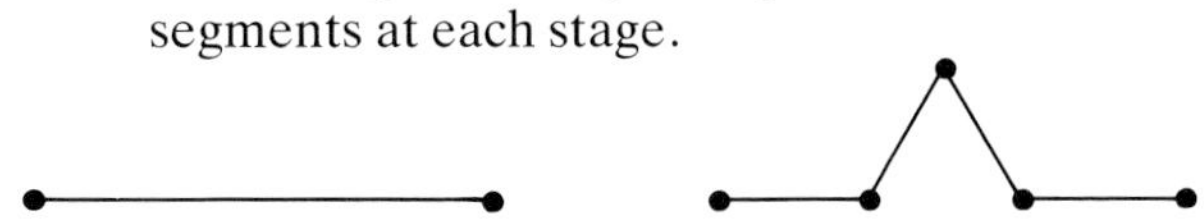

(ii) the length of each new segment is ⅓ of the length of the segment from which it is formed.

(iii) each new equilateral triangle added to the snowflake is ⅑ of the area of the triangle to which it is added.

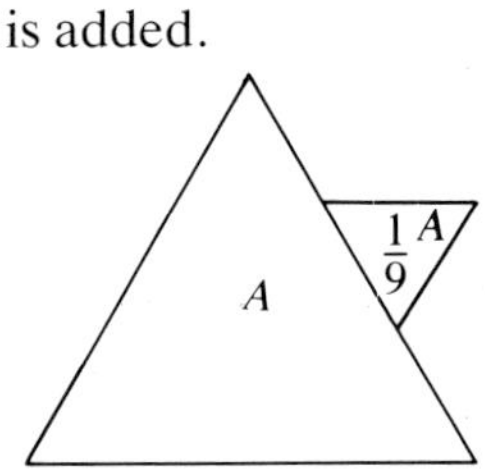

To determine the length of the curve
Initially there are 3 segments of length l. Applying the procedure *once* produces 4×3 segments of length $\frac{l}{3}$,

twice produces $4 \times 4 \times 3$ segments of length $\frac{1}{3} \times \frac{l}{3}$ etc. After applying the procedure n times there are $4^n \times 3$ segments of length $(\frac{1}{3})^n l$, i.e. the length of the snowflake curve is given by

$$3 \times \left(\frac{4}{3}\right)^n l$$

Thus if the process is applied indefinitely, i.e. $n \to \infty$ then the length of the curve increases without limit.

To determine the area enclosed by the snowflake

Each time the procedure is applied the area increases by the number of new triangles added. One new triangle is added for each segment. Thus the nth time the procedure is applied the increase in area is given by

$$(4^{n-1} \times 3) \times \left(\frac{1}{9^n} A\right)$$

and the total area is:

$$A + 3 \times \left(\frac{1}{9} A\right) + (4 \times 3) \times \left(\frac{1}{9^2} A\right)$$

$$+ (4^2 \times 3) \times \left(\frac{1}{9^3} A\right) + \ldots + (4^{n-1} \times 3) \times \left(\frac{1}{9^n} A\right)$$

$$= A\left(1 + \frac{3}{9}\left(1 + \frac{4}{9} + \left(\frac{4}{9}\right)^2 + \ldots + \left(\frac{4}{9}\right)^{n-1}\right)\right)$$

Thus the area is determined by summing a geometric series of common raio 4/9. If the process is applied indefinitely, i.e. $n \to \infty$, the area converges to

$$A\left[1 + \frac{3}{9} \times \frac{1}{\left(1 - \frac{4}{9}\right)}\right] = 1.6A$$

In conclusion, although the length of the snowflake increases without limit, the area it encloses never exceeds 1.6 times the area of the original triangle.

An extension of this problem is to use a square, rather than equilateral triangle as the basic shape, as illustrated in below.

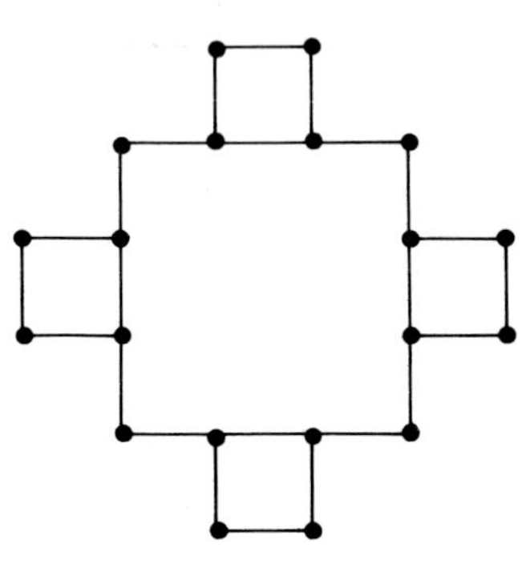

Similar results are obtained for the length of the curve and the area it encloses.

Solution for the square

In the case of the square, each side of a square produces five new segments each time the procedure is applied, as shown here.

As before the area of each square added is $\frac{1}{9}$th the area of the square to which it is added. Thus after applying the procedure n times the total length is

$$4 \times 5^n \times \left(\frac{l}{3^n}\right) = 4 \times \left(\frac{5}{3}\right)^n l$$

and the total area enclosed is

$$= A + 4 \times \left(\frac{1}{9} A\right) + 4 \times 5 \times \left(\frac{1}{9^2} A\right)$$

$$+ 4 \times 5^2 \times \left(\frac{1}{9^3} A\right) + \ldots + (4 \times 5^{n-1}) \times \left(\frac{1}{9^n} A\right)$$

$$= A\left(1 + \frac{4}{9}\left(1 + \frac{5}{9} + \left(\frac{5}{9}\right)^2 + \ldots + \left(\frac{5}{9}\right)^{n-1}\right)\right)$$

Thus if the procedure is applied indefinitely, i.e. $n \to \infty$, the total length increases without limit, and the area enclosed converges to

$$A\left[1 + \frac{4}{9} \times \frac{1}{\left(1 - \frac{5}{9}\right)}\right] = 2A$$

Investigation 18: Into the fourth dimension and beyond

Here are the eight cubes which make up a hypercube.

Here are the six squares which make up a cube.

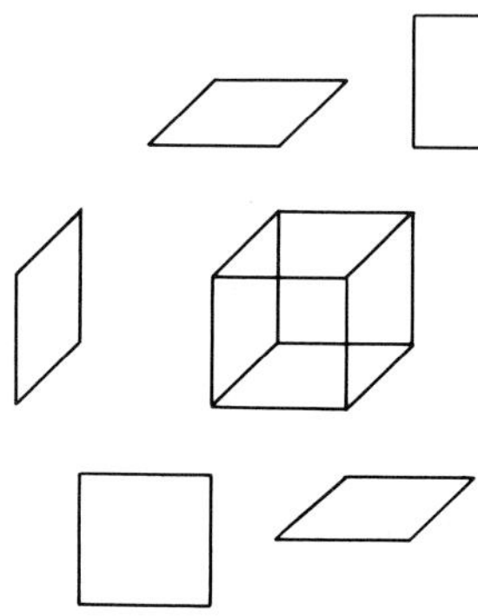

Let students give their own names to shapes beyond the hypercube (which is also known as a tesseract).

Using notation taken from H.S.M. Coxeter's book *Regular Polytopes*, $\{p,q,r\}$ is a 4 dimensional polytope with r solids to an edge where each solid is made of polygons, q of which meet at a vertex. Coxeter uses the symbols α, β and γ so that the cube is $\gamma_3 = \{4,3\}$ (i.e. 3 4-gons to a vertex), then $\gamma_3 = \{4,3,3\}$ is the hypercube (i.e. 3 cubes, or $\{4,3\}$s, to an edge), $\gamma_5 = \{4,3,3,3\}$ is the 5 dimensional analogue of a cube (i.e. 3 hypercubes to a face), and so on. Similarly the tetrahedron is $\alpha_3 = \{3,3\}$ (i.e. 3 3-gons to a vertex), the simplex in 4 dimensions is $\alpha_4 = \{3,3,3\}$ (i.e. 3 tetrahedra to an edge, each tetrahedron being 3 3-gons to a vertex), and in 5 dimensions it is $\alpha_5 = \{3,3,3,3\}$ (i.e. 3 α_4's to a face), and so on. The octahedron is $\beta_3 = \{3,4\}$ (i.e. 4 3-gons to a vertex), then the 16-cell is $\beta_4 = \{3,3,4\}$ (with 4 tetrahedra or α_3's to an edge), and the analogue in 5 dimensions is $\beta_5 = \{3,3,3,4\}$ (with 3 α_3's to an edge and 4 α_4's to a face), and so on.

Coxeter uses F_0 for the number of vertices, F_1 for the number of edges, F_2 for the number of faces, F_3 for the number of solids etc. By considering how the number of n-gon faces meeting at a vertex and so on, relates to the number of faces, vertices, edges, etc, and using the Euler-Poincaré formula,

$$\sum_{r=0}^{n} \left((-1)^r F_r\right) = 1$$

it is possible to work out F_0 to F_n for any n dimensional polytope given its definition in $\{a_1, a_2, a_3, \ldots, a_{n-1}\}$ form. Using this it is possible to build up a picture of the cube family without relying on sequences where we only have the first three terms (which is how many students may do it). In practice it is easier to find F_0 by considering how each polytope is built up from the last. For the γ_n polytopes F_0 doubles each time n is increased by 1. For the α_n polytopes F_0 increases by 1 each time n is increased. The Euler-Poincaré formula is then a useful check for each line of the table.

For example for γ_2 (a square):

$F_0 = 4$ (4 vertices)

$F_1 = 4$ (4 edges)

$F_2 = 1$ (1 face)

So for γ_3:

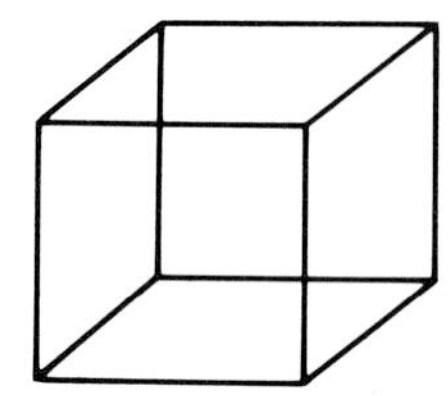

F_0 for $\gamma_3 = 2 \times F_0$ for $\gamma_2 = 8$ (vertices are doubled)

F_1 for $\gamma_3 = 2 \times F_1$ for $\gamma_2 + F_0$ for $\gamma_2 = 12$ (edges are doubled and each vertex of a square generates an edge of γ_3)

F_2 for $\gamma_3 = 2 \times F_2$ for $\gamma_2 + F_1$ for $\gamma_2 = 6$ (faces are doubled and each edge of a square generates a face of γ_3)

F_3 for $\gamma_3 = 1$ (one solid)

As a check

$$\sum_{r=0}^{3} \left[(-1)^r F_r\right] = F_0 - F_1 + F_2 - F_3 = 8 - 12 + 6 - 1 = 1$$

For γ_4:

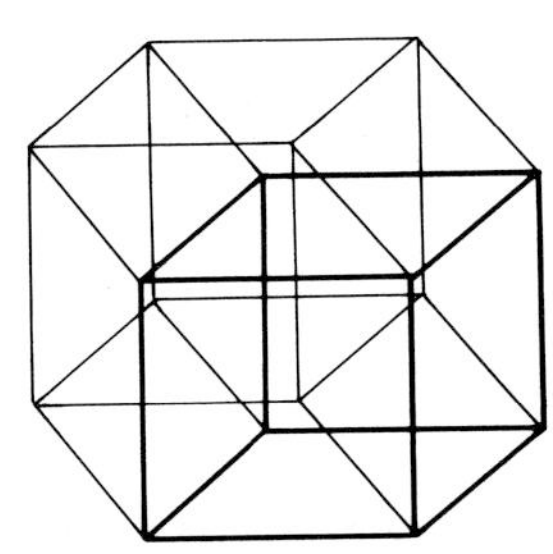

F_0 for $\gamma_4 = 2 \times F_0$ for $\gamma_3 = 16$
F_1 for $\gamma_4 = 2 \times F_1$ for $\gamma_3 + F_0$ for $\gamma_3 = 32$
F_2 for $\gamma_4 = 2 \times F_2$ for $\gamma_3 + F_1$ for $\gamma_3 = 24$
F_3 for $\gamma_4 = 2 \times F_3$ for $\gamma_3 + F_2$ for $\gamma_3 = 8$
F_4 for $\gamma_4 = 1$

As a check

$$\sum_{r=0}^{4} \left[(-1)^r F_r\right] = F_0 - F_1 + F_2 - F_3 + F_4 = 16 - 32 + 24 - 8 + 1 = 1$$

Table for γ_n

n	F_0	F_1	F_2	F_3	F_4	F_5	F_6	F_7
1	2	1						
2	4	4	1					
3	8	12	6	1				
4	16	32	24	8	1			
5	32	80	80	40	10	1		
6	64	192	240	160	60	12	1	
7	128	448	672	560	280	84	14	1

It soon becomes obvious that for an n dimensional cube,

$\gamma_n = \{4,3,3,\ldots,3),$

$F_0 = 2^n$

$$F_1 = 2^n \left(\frac{n}{2}\right)$$

$$F_2 = 2^n \left(\frac{n}{2}\right)\left(\frac{n-1}{4}\right)$$

$$F_3 = 2^n \left(\frac{n}{2}\right)\left(\frac{n-1}{4}\right)\left(\frac{n-2}{6}\right)$$

.
.
.

$$F_m = 2^n \left(\frac{n}{2}\right)\left(\frac{n-1}{4}\right)\left(\frac{n-2}{6}\right)\cdots\left(\frac{n-m+1}{2m}\right)$$

$$\text{i.e. } F_{m+1} = F_m \left(\frac{n-m+1}{2m}\right)$$

Similarly for the tetrahedra polytopes:

LINE

Add a point in a perpendicular direction and join up the points.

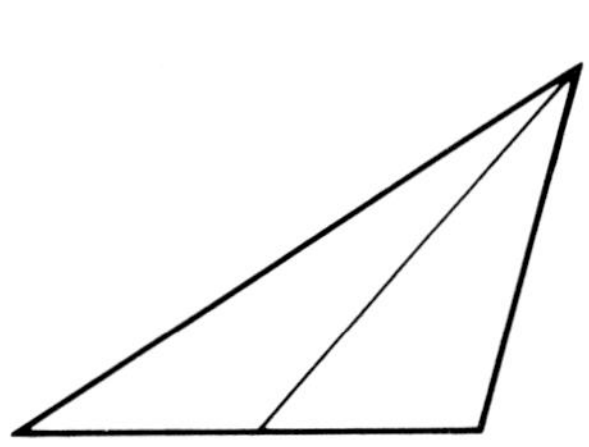

TRIANGLE

Add a point in a perpendicular direction to the first direction, and join up the points.

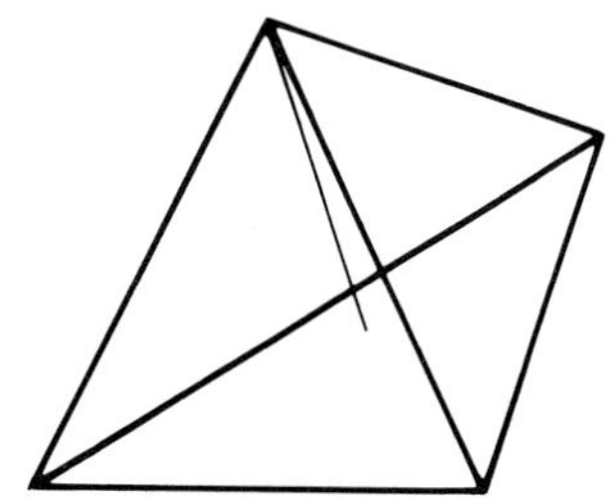

TETRAHEDRON

Add a point in a perpendicular direction to the first two directions and join up points.

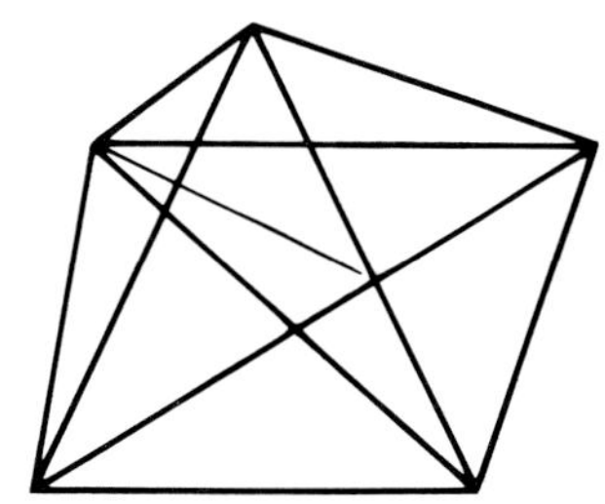

4D SIMPLEX

Add a point in a perpendicular direction to the first three directions and join up points

Table for α_n

n	F_0	F_1	F_2	F_3	F_4	F_5	F_6	F_7
1	2	1						
2	3	3	1					
3	4	6	4	1				
4	5	10	10	5	1			
5	6	15	20	15	6	1		
6	7	21	35	35	21	7	1	
7	8	28	56	70	56	28	8	1

$$\text{and} \quad F_{m+1} = F_m \left(\frac{n-m+1}{m+1}\right)$$

The pattern of Pascal's triangle springs to the eye.

Extensions

(a) Draw a 1D net for a square
a 2D net for a cube
a 3D net for a hypercube.

(b) How can you write coordinates in more than 3 dimensions?

(c) Define the distance between points in 4,5,6, . . . n dimensions, and the midpoint of two points.

(d) Define a hypervolume to measure hypercubes in.

(e) Find the equations of lines and planes in 4 dimensions using vectors or coordinate geometry. Try to go one step further on.

(f) What is the equation of a

(i) circle in 2D
(ii) sphere in 3D
(iii) hypersphere in 4D?

What about n dimensions?

(g) Can you make 3 dimensional projections of any 4 dimensional polytopes? Are there any 4 dimensional regular polytopes other than the two you have found so far?

(h) What 5 dimensional, 6 dimensional, n dimensional regular polytopes are there?

(i) For 3 dimensional polyhedra, $V-E+F=2$. What about 2 dimensional polygons or 4 dimensional polytopes?

Notes on the extensions

(a) 1D net for a square

2D net for a cube

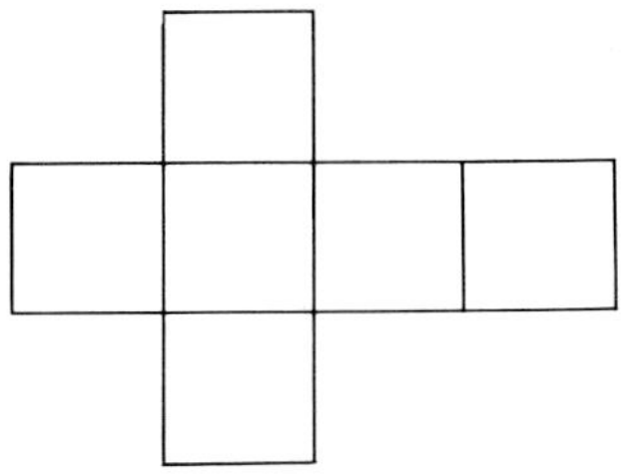

3D net for a hypercube

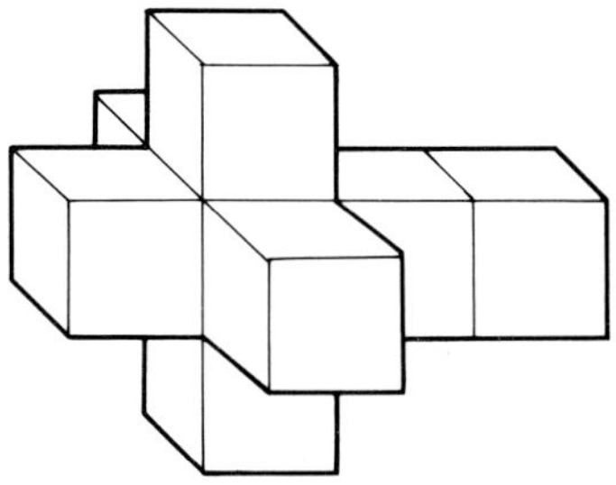

Salvador Dali's painting 'Corpus Hypercubus' depicts Christ on a 3D net of a hypercube. Robert Heinlein has written a short story called 'And He Built a Crooked House' about a house built in the form of a 3D net of a hypercube (which folds up trapping the occupant inside forever!)

(b) Coordinates can easily be written in any number of dimensions $(x_1, x_2, x_3, \ldots x_n)$.

(c) Pythagoras' theorem can be extended into higher dimensions. The distance between (x_1, x_2, x_3) and (y_1, y_2, y_3) is

$$\sqrt{(x_1-y_1)^2+(x_2-y_2)^2+(x_3-y_3)^2}$$

The distance between $(x_1, x_2, x_3, \ldots, x_n)$ and $(y_1, y_2, y_3, \ldots, y_n)$ is

$$\sqrt{(x_1-y_1)^2+(x_2-y_2)^2+(x_3-y_3)^2+\ldots+(x_n-y_n)^2}$$

This will measure the distance between opposite vertices of an n-dimensional cube.

(d) A hypervolume could be measured in unit hypercubes, $1 \times 1 \times 1 \times 1$ cm. So a hypercube with edges 2 cm long would have a hypervolume of 2^4 cm^4.

(e) Any line in any number of dimensions can be written as $\mathbf{r} = \mathbf{a} + \lambda\mathbf{b}$. Cartesian equations can be deduced from this for 4 dimensions, but they are not very pretty. Similarly a plane can be defined in 4D using $\mathbf{r} = \mathbf{a} + \lambda\mathbf{b} + \mu\mathbf{c}$ but it is no longer sufficient to find a vector normal to the plane to define it in $\mathbf{r}\cdot\mathbf{n} = k$ form since there will be an infinite number of non-parallel planes which are perpendicular to any given vector (just as there are an infinite number of non-parallel lines in 3 dimensions which are perpendicular to any given vector).

(f) **(i)** circle in 2D $(x-a)^2+(y-b)^2=r^2$
(ii) Sphere in 3D $(x-a)^2+(y-b)^2+(z-c)^2=r^2$
(iii) Hypersphere in 4D
$(x_1-a)^2+(x_2-b)^2+(x_3-c)^2+(x_4-d)^2=r^2$

(g) 3D projections of the 4D simplex and hypercube are easy to make from straws. The hypercube can be made from two interlocking cubes of the same size with vertices joined. It can also be made as a smaller cube inside a larger cube with vertices joined. (The 'inside' cube is smaller due to perspective!)

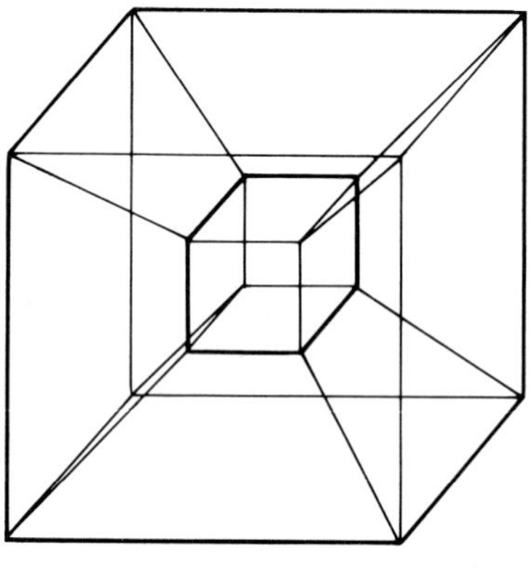

There are 6 regular polytopes in 4D. They are shown in the table on the next page. Students would probably need to go to text books to find this out.

Polytope	Vertices	Edges	Faces	Solids	Type of solid
simplex	5	10	10	5	tetrahedra
hypercube	16	32	24	8	cubes
16-cell	8	24	32	16	tetrahedra
24-cell	24	96	96	24	octahedra
120-cell	600	1200	720	120	dodecahedra
600-cell	120	720	1200	600	tetrahedra

(h) In all higher dimensions (5,6, upwards) there are only 3 regular polytopes. These three are analogous to the tetrahedron, the cube and the octahedron. In 2 dimensions, of course, there is an infinite number of regular polygons. Books such as Edwin Abbotts' Flatland and A.K. Dewdney's The Planiverse are excellent introductions to thought on 2 dimensions, which leads into thought on higher dimensions.

Investigation 19: Continued fractions

Any rational number can be written as a finite continued fraction.

Eg. $\frac{6}{5} = 1 + \frac{1}{5} = [1;5]$

$$\frac{3}{8} = \frac{1}{\left[\frac{8}{3}\right]} = \frac{1}{2+\frac{2}{3}} = \frac{1}{2+\frac{1}{\left[\frac{3}{2}\right]}}$$

$$= \frac{1}{2+\frac{1}{1+\frac{1}{2}}} = [0;2,1,2]$$

etc.

$$[1;1,2,3] = 1 + \frac{1}{1+\frac{1}{2+\frac{1}{3}}} = 1 + \frac{1}{1+\frac{3}{7}}$$

$$= 1 + \frac{7}{10}$$

$$= \frac{17}{10}$$

$[0;2,1] = 0 + \frac{1}{2+\frac{1}{1}} = \frac{1}{3}$ so why not say [0;3]?

In fact, it is never necessary for a continued fraction to end with a one.

$$[1;\overline{1,2}] = 1 + \frac{1}{1+\frac{1}{2+\frac{1}{1+\frac{1}{2\ \ldots}}}} = x$$

$$= 1 + \frac{1}{1+\frac{1}{1+x}} = x$$

So $x = 1 + \frac{1}{\frac{1+x+1}{1+x}} = 1 + \frac{1+x}{2+x}$

$$x = \frac{2+x+1+x}{2+x}$$

$$(2+x)x = 3 + 2x$$

$$x^2 + 2x - 2x - 3 = 0$$
$$x^2 = 3$$
$$x = \sqrt{3}$$

This must be positive since we have been adding positive amounts together. All continued fractions, as defined in the original statement, are positive.

Some students may wish to investigate what happens if a new type of fraction is defined with subtractions instead of additions, or alternate additions and subtractions.

Once you have found $\sqrt{3} = [1;\overline{1,2}]$, it is natural to wonder about $\sqrt{2}$, $\sqrt{5}$ etc.

$$\sqrt{2} = 1 + (\sqrt{2}-1)$$

$$= 1 + \cfrac{1}{\cfrac{1}{\sqrt{2}-1}}$$

$$= 1 + \cfrac{1}{\cfrac{\sqrt{2}+1}{1}} = 1 + \frac{1}{1+\sqrt{2}}$$

So $$\sqrt{2} = 1 + \frac{1}{1+\sqrt{2}}$$

$$1 + \cfrac{1}{1 + \ 1 + \cfrac{1}{1+\sqrt{2}}} = 1 + \cfrac{1}{2 + \cfrac{1}{1+\sqrt{2}}}$$

$$= 1 + \cfrac{1}{2 + \cfrac{1}{1 + \ 1 + \cfrac{1}{2 + \cfrac{1}{1+\sqrt{2}}}}}$$

$$= 1 + \cfrac{1}{2 + \cfrac{1}{2 + \cfrac{1}{2 + \cfrac{1}{\cdots\cdots}}}} = [1;\overline{2}]$$

Similarly it can be found that:

$\sqrt{2} = [1;\overline{2}]$
$\sqrt{3} = [1,\overline{1,2}]$
$\sqrt{5} = [2,\overline{4}]$
$\sqrt{10} = [3;\overline{6}]$
$\sqrt{11} = [3;\overline{3,6}]$
$\sqrt{12} = [3;\overline{2,6}]$
$\sqrt{13} = [3;\overline{1,1,1,1,6}]$
$\sqrt{6} = [2,\overline{2,4}]$
$\sqrt{7} = [2,\overline{1,1,1,4}]$
$\sqrt{8} = [2,\overline{1,4}]$
$\sqrt{14} = [3;\overline{1,2,1,6}]$
$\sqrt{15} = [3;\overline{1,6}]$
$\sqrt{17}$ is nice $[4;\overline{8}]$
$\sqrt{26} = [5;\overline{10}]$
$\sqrt{x^2+1} = [x;\overline{2x}]$
$\sqrt{x^2-1} = [x-1];\overline{1, 2[x-1]}$
$\sqrt{x^2+2} = [x;\overline{x,2x}]$

$[1;\overline{1}]$ is also interesting continued fraction

$$[1;\overline{1}] = 1 + \cfrac{1}{1 + \cfrac{1}{1 + \cfrac{1}{1 + \cdots\cdots\cdots\cdots}}}$$

$$x = 1 + \frac{1}{x} = \frac{x+1}{x}$$

$$x^2 = x + 1$$

$$x^2 - x - 1 = 0$$

$$x = \frac{1 \pm \sqrt{5}}{2}, \ x > 0 \ \text{so}$$

$$x = \frac{1+\sqrt{5}}{2}$$

which is the **golden ratio** – that to which the numbers in Fibonacci's sequence tend if divided:

$$\frac{1}{1}, \frac{2}{1}, \frac{3}{2}, \frac{5}{3}, \frac{8}{5},$$

$$\frac{13}{8}, \frac{21}{13}, \frac{34}{21}, \frac{55}{34}, \text{etc.}$$

Since continued fractions are positive, only positive real roots of quadratics can be found in continued fraction form.

All positive rational roots can be written as finite continued fractions.

Positive irrational roots can be rewritten using the methods above. For example:

$$\frac{3-\sqrt{3}}{2}$$

$$\frac{3-\sqrt{3}}{2} = \cfrac{1}{\cfrac{2}{3-\sqrt{3}}} = \cfrac{1}{\cfrac{2(3+\sqrt{3})}{9-3}} = \cfrac{1}{\cfrac{6+2\sqrt{3}}{6}}$$

$$= \frac{1}{1+\frac{\sqrt{3}}{3}} = \frac{1}{1+\frac{1}{\sqrt{3}}}$$

$$= \cfrac{1}{1+\cfrac{1}{1+\cfrac{1}{1+\cfrac{1}{2+\cfrac{1}{1+\cfrac{1}{2+\ldots}}}}}} \quad \text{since } \sqrt{3} = [1;\overline{1,2}]$$

$$= [0;1,1,\overline{1,2}]$$

Investigation 20: Functions defined by integrals

The purpose of this investigation is to encourage students to realise that functions can be defined in different ways from a formal rule such as $F(x) = \sin x$. A common definition of a function is in terms of an integral, for example

$$\int_0^x e^{-t^2} \mathrm{d}t$$

occurs in the normal distribution in statistics.

Encourage students to explore standard functions, and functions that they cannot integrate analytically.

Discuss the domain and range of their new functions.

$f(t)$	$F(x)$
t^n	$\frac{x}{n+1}$
$\sin t$	$1 - \cos x$
$\cos t$	$\sin x$
e^t	$e^x - 1$
$\begin{cases} 1 & 0 \leqslant t \leqslant 1 \\ 0 & t > 1 \end{cases}$	$\begin{cases} x & 0 \leqslant x \leqslant 1 \\ 1 & x > 1 \end{cases}$
$\begin{cases} t & 0 \leqslant t \leqslant 2 \\ 0 & t > 2 \end{cases}$	$\begin{cases} \frac{x^2}{2} & 0 \leqslant x \leqslant 2 \\ 2 & x > 2 \end{cases}$

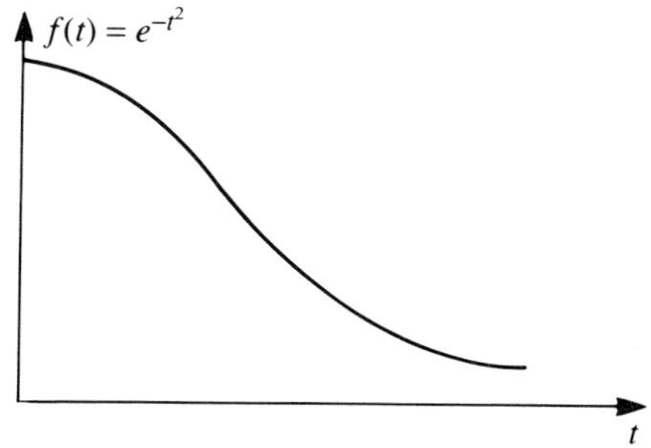

For $f(t) = e^{-t^2}$ values of $F(x)$ are:

x	0	1	2	3
$F(x)$	0	0.7468	0.8821	0.8862

A graph of $F(x)$ is shown in the figure below.

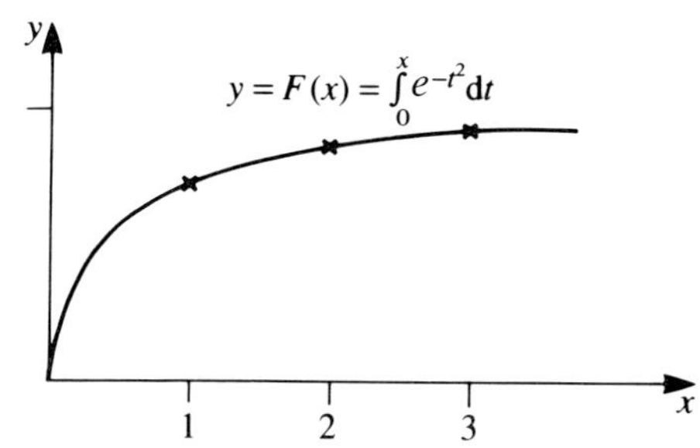